8 50
0, L.

D1505715

LIONS SHARE
The Story of a Serengeti Pride

LIONS SHARE

The Story of
a Serengeti Pride

Written by
JEANNETTE HANBY

Illustrated by
DAVID BYGOTT

Foreword by Jane Goodall

Houghton Mifflin Company   Boston   1982

Library of Congress Cataloging in Publication Data

Hanby, Jeannette.
    Lions share.

    1. Lions—Behavior.   2. Mammals—Behavior.   3. Mam-
mals—Tanzania—Serengeti Plain—Behavior.   4. Serengeti
Plain (Tanzania)   I. Title.
QL737.C23H35   599.74′428   82-3065
ISBN 0-395-32043-7   AACR2

Printed in the United States of America

V 10 9 8 7 6 5 4 3 2 1

CONTENTS

FOREWORD *by Jane Goodall*

One day several kinds of animals joined the lion in a hunt. After the kill, when the prey was to be divided, the lion claimed one quarter as his right, one quarter for his superior courage, one quarter for his dam and cubs – "and as for the fourth quarter, let who will dispute it with me". The other beasts, awed by his frown, yielded and silently withdrew.

This story, one of Aesop's fables, presents the traditional image of the lion as King of Beasts. It is not difficult to guess why he was so named. One only has to watch him as he stands, the wind slightly ruffling his magnificent mane, his proud, cold, golden eyes looking out over his African kingdom; one only has to hear, resounding over the plains, his awe-inspiring series of roars, challenging any intruder. His bearing, at such times, is utterly majestic. No wonder the lion – rampant, passant, or salient – was chosen as a heraldic emblem to enhance the royalty of kings, princes and emperors throughout the ages. No wonder that almost every visitor to the African game park demands, first and foremost, to see the lions. For lions have captured the imagination and stirred the emotions of countless generations.

I met my first lions in Africa when I was out for an evening walk in the Olduvai Gorge, on the Serengeti. I was steeped in the jungle lore of Mowgli and Tarzan and was convinced that no wild animal would hurt me unless I was afraid. This, in fact, is not always true, but on that occasion the two young males were merely curious, and after following me a short way, wandered off about their business.

Since then I have watched many lions. I have marvelled at the controlled power in the lithe bodies of hunting females, and smiled at the clumsy efforts of small cubs chasing butterflies or their elders' waving tail tips. I have observed them mating and fighting and playing. And once I saw the shocking ferocity with which a big male killed a hyena which did not run fast enough. But mostly the lions I saw were resting or sleeping, for this is how they spend the greater part of every day; the really interesting behaviour usually takes place at night.

It needs people with special qualifications to carry out a really good study of lions; people with endless patience, persistence and an unflagging interest in the behaviour of their subjects. I would go further and say, at the risk of raised eyebrows from some of my colleagues, that the observers need to develop a degree of empathy with the lions themselves. All these qualities, together with a love of the African bush and the ability to live without the comforts of civilization, are possessed in good measure by the authors of this unique book.

I have known David for over ten years, since he came to work for me at Gombe collecting data on chimpanzee behaviour. He went on to do a further year there on his own research project, male dominance, for a PhD degree at Cambridge University. Before he left, David developed a deep, intuitive understanding of the chimpanzees. At least, I thought so but then I was biased – for so often his interpretations happened to correspond with my own!

David met and fell in love with Jeannette at Cambridge where she was working on a research project on monkeys. Jeannette is as lively and outgoing as David is quiet and self-contained. Both are hard workers, keen observers, excellent field naturalists and dedicated conservationists. They make an unusually talented team, for as well as being scientists of unquestioned integrity, both are artists – Jeannette with words and David with pictures.

Jeannette has written the text, describing the observations that both she and David made, and expressing the thoughts and ideas which they talked about together, often no doubt as they sat in their Land-Rover while the lions slept away the hot day outside. Her prose does justice not only to the lions themselves but also to the country in which they live. I know and love the Serengeti for I lived there for over two years; reading *Lions Share* made me homesick, all over again, for the wide open spaces, the clear air, the ever-changing panorama of the sun-drenched, wind-swept plains, that sense of freedom and the eternal nature of things.

David's drawings provide an added dimension, particularly important for readers who have never been to the Serengeti. Most wild life books are illustrated with photographs, and often these pictures are selected for their *quality*. David has been able to select the most vivid and important *events* of the study to illuminate this book; moreover, his long hours of familiarity with lions and their world have enabled him to express certain moods that would be difficult to capture with film.

This book presents a wealth of information about lions, much of it that is quite new, presented in a manner which almost forces one to become deeply involved in the lives of the various characters. Nine females, pushed out into new, often hostile territory, fleeing from enemies, threatened by drought – can they possibly survive? What is a lioness to do when practically all the prey she catches is taken from her by her male protector? If she leaves him she will be in danger from enemies. If she stays she will be in danger of starvation. What is the solution for the mother who is forced to give birth in the middle of a dry semi-desert where there is no water and precious little food?

The book underlines the tremendous importance of long term study if one is to begin to understand the complex behaviour and society of long-lived creatures such as lions. The Serengeti lion study was begun by George Schaller in 1966, taken over by Brian Bertram, who was followed by the Bygotts. They, in turn, handed over to the Packers. It is a very real tragedy that, since the Packers left in 1980, there has been no scientist watching the Serengeti lions. The story that emerges in this book, the story of one pride, is immeasurably richer because, by checking with previous records, the Bygotts were able to establish the origins of the Sametu females and their relationship to the five great males who terrorized them for so long.

David and Jeannette have kept themselves entirely out of this story. As one reads, engrossed in the drama of the lions, it is easy to forget the sheer hard work that is going on behind the scenes, the long hours of detailed note-taking and checking of records, the days spent sitting and waiting while the lions slept, the arduous bumping over rough terrain, the dust and the heat and the flies. We should, just occasionally, pause to remember the humans who documented this unique account. The book is based entirely on fact although, as the authors point out in their Introduction, they have spiced the facts with their interpretations from time to time. After nearly four years of living with lions it would be unfortunate if they could not provide us with insights into the meaning of the behaviour they describe – and a poorer book.

I would like to thank Jeannette and David for *Lions Share*, not only because I have found it utterly fascinating and learned a great deal from it, but because books such as this are of major importance to the cause of conservation. It becomes a matter of concern to all of us that the magnificent Serengeti plains should be preserved, for all time, so that the Sametu Pride, and its descendents, can live out their lives in freedom.

INTRODUCTION

Lions share. They share food, mates, cubs and a homeland in which they live and hunt. Lions who share are called a pride. This is a true story about a pride living on the open plains of Serengeti, particularly about Sonara, one of its members. Sonara is one of nine females in the Sametu pride; she embodies the experiences, affection, cooperation, skill and playfulness of the pride. The book is also about the Sametu pride's neighbours; other settled prides, wandering nomads and the rich diversity of animals – from aardvarks to zebras – that share the plains with the lions.

When we first saw the Sametu pride we were squeezed together in the back seat of a Super-cub aircraft. Since dawn we had been systematically searching for lions over the heart of Serengeti National Park, Tanzania. Early on, we had been rewarded with glimpses of lions perched on rocks, lying by waterholes or lazing in the shade of acacia trees, but then this wooded country gave way to open grassland, a vast plain with few landmarks or lions. In the heat of midday we wearily scanned the expanse of threadbare brown carpet that stretched below us.

Then we saw them – nine tiny toy-like creatures – lying as if dropped accidentally in the middle of nowhere. We were surprised to find so large a group on the plains; they were far from any water or shade, and prey animals were scarce. Immediately after the flight we drove out in search of them. It was late afternoon when we reached the group, still sprawling on a low ridge overlooking the Sametu valley. The lions were sleek and healthy and totally unafraid of our Land-Rover. All the females were just mature, about three and a half years old. None of them matched any lions in our identity file, so we began the task of drawing up a new card for each individual.

11

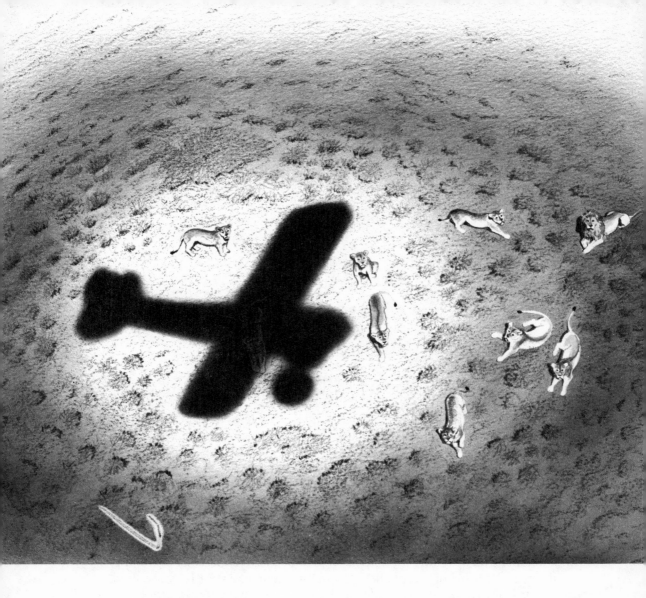

On the front of the card we drew an outline of the face, sketching any ear notches which the lion had acquired during mealtime squabbles or more serious fights. We also drew the unique pattern of black whisker-spots on both sides of the muzzle; like a fingerprint, this stays the same from birth to death. We noted other useful marks such as missing teeth or tail-tips. In the evening light, we took close-up photos of each side of the lion's face, to be pasted on the back of the card when developed. To complete the task we gave each lion a name. Our system was to name the group after the area in which we found it, in this case, Sametu. Each individual was given a name beginning with the letter "S". (You will find a list of the names with their pronunciations and meanings, at the end of this Introduction.)

We were not equipped to follow the lions that night and they slipped away from us at dusk. Our newly named Sonara came close, sniffed our tyres and peered at us before bounding off playfully to join the others already setting out on their evening hunt. They were a captivating group and a real mystery: Where had they come from? Why were they so tame? Above all, how could they survive amidst such barren emptiness?

Discovering the origins of the Sametu pride involved some detective work. Fortunately, we had access to records gathered by our predecessors on the Lion Project. From these we learned that sixteen large cubs had disappeared from the Masai pride. They had been just two years old, barely able to support themselves; their fate was unknown. Dr Brian Bertram had been studying the lions at that time and had managed to photograph a few of the youngsters before they had vanished. We compared their pictures with those of our new-found Sametu pride. Some matched perfectly. Thus their origins could be traced; the nine adult Sametu females were the survivors of the many cubs born to the Masai pride in 1971. This also explained why our mystery lions were so tame. They had been familiar with cars from an early age, the Masai pride being one of the most popular with tourists and also studied by scientists since Dr George Schaller initiated the Lion Project in 1966.

Discovering how the Sametu pride could live on the plains throughout the year took up an exciting portion of our four-year study of Serengeti lions. The plains are not hospitable to lions except during the rainy season which lasts roughly from November through May. Then the plains are a welcome sight – a vast green lawn jewelled with sparkling lakes and swarming multitudes of animals. But when the plains dry up, between June and October, the plains look like a desert – a windy expanse of parched grass and fine dust. This great contrast in seasons means that lions usually come onto the plains only during the rains. Most lions keep to the wooded areas of Serengeti at all times of year where water, food and prey are always available. During our study we found a few prides living a precarious existence on the plains even in dry season. This implied that conditions had changed somewhat, for the better from the lions' point of view.

We checked the ecological records kept by the Serengeti Research Institute and found that the dry seasons had not been quite so dry in recent years. Thus there was more rainfall, grass, grazers and consequently, a few hardy lions. The slight increase in dry season rainfall didn't mean that life on the plains was luxurious for the lions, it meant only that they could try to reside there all year round. The Sametu pride was the largest of the pioneering groups and this book is the story of their struggle for survival.

Just as we have tried to keep out of the lions' lives, we have tried to keep out of their story. Most "animal books" are really about people, real people who keep or study animals, or people disguised as animals (endowing animals with human thoughts and personalities, called "anthropomorphism", is something we have tried to avoid). In this book, the lions are the stars of the story.

We are trying to give you the lions' view of their world, without interference or judgment, but with some degree of interpretation. Thus we describe what we saw the lions do and lace the action together with insights drawn from our own

– and other people's – research. Our indebtedness to our fellow scientists, especially those at the Serengeti Research Institute, is immense. We have freely utilized information from the long-term Lion Project and from the team effort to understand the Serengeti ecosystem as a whole. Years of many people's work have been spun into the strands that give a wider validity to the lions' story.

The book has three international themes: first, the lions' story; second, their social structure and behaviour in general; third, the ecology of the Serengeti plains.

The story of the Sametu pride, its neighbours and enemies, is an actual case history that documents the way a particular group of lions lives in Serengeti. These lions are not "average" but exemplify how lions live in a place with striking seasonal changes in food supply. The Sametu pride is exceptional in that it is relatively large and composed of adult females all of one age-set. It is a pioneering pride, having to establish a territory, learn to rear cubs and hunt in a new and harsh environment. The other lions demonstrate the typical variation seen in lion society – from solitary nomads to bands of males that "own" groups of females. There are four major groups; the Sametu sisters; the Loliondo males; Kesho and Kali and the Nomad Trio – and two minor groups – the Masai pride and the Boma pride. The major groups are all composed of relatives except for the Nomad Trio. The words "sister" and "brother" are used throughout the book to refer to closely related lions (full or half-siblings, cousins) that are roughly the same age and from the same natal pride (they are also called age-mates, or collectively, an age-set).

The ranges of the various prides can be located on the maps at the endpapers. Spellings, pronunciations and meanings of individual lion names can be found in the Appendix. Also in the Appendix is a condensed chronology, giving a chapter by chapter summary of the important events that happened in the four major lion groups.

Because the Serengeti Lion Project has been long term, we know that the behaviour and relationships of the lions in this particular story are representative of lions elsewhere. We emphasize and explain certain aspects of the lions' actual experiences in order to illustrate general principles of social organization and behaviour. For example, the first two chapters lay out the basics of lion society: the structure of the pride; nomadic lions; the different roles of the sexes, territoriality etc. Chapter one focuses on the behaviour of females; chapter two gives details of male behaviour.

Subsequent chapters deal with hunting, scavenging, exploration, ranging, mating, cub bearing, rearing and development, defence, disease, survival and ways of coping with an environment that changes seasonally and socially. What we hope emerges is a fairly complete picture of how wild lions live.

The lions are the stars but their story cannot be properly told without an account of the kingdom in which they live. By no means do we have a complete picture of how this ecosystem works, but we do have an appreciation of its complexity and beauty, and we can at least sketch out a framework. The third chapter describes how the basic elements – sun, soil, water, grass, animals –

work together to promote one of the most productive of all earthly environments. Each chapter in the book describes various aspects of the Serengeti plains; collectively they are meant to convey the ebb and flow of the seasons that so markedly affect all life there.

The animals that share the Serengeti with the lions have their place in the tale too. Resident or migrant, nocturnal or diurnal; flying, creeping, running or burrowing, predator or prey, there are so many different creatures, their societies and behaviour so interesting that it is impossible to do them justice. Over one hundred different species are mentioned or illustrated, and the drawings of the sometimes strange but always wonderful birds, beasts and vegetation are worth several times more than the few lines in the text. What we hope to communicate is a sense of the diversity and the dynamic quality of this incredibly rich place.

We have tried to convey something of the atmosphere of Serengeti by illustrating the story with drawings. This medium allows us to portray events and places that would have been difficult or impossible to photograph well. We hope that the illustrations are as enjoyable to look at as they were to prepare.

The Masai people gave Serengeti its name, which means "wide open space", a place of vast skies and far horizons. We hope that through this book you will be able to share that sense of wildness and freedom that the Serengeti evokes. We hope too, that you will come to know the Serengeti as we did, not as the bare brown carpet we saw at first, but as a rich tapestry of interwoven colours through which the lions wander like golden strands.

LIONS SHARE

Shiba

Safi

Sarabi

Sega

Sukari

Siku

Salama

Sonara

Swala

CHAPTER ONE

Sisters
(Late dry season, 1st Year)

Sonara awakened slowly as the shade crept away to the other side of her rock. The hot afternoon sun glared at her white belly and she could no longer sleep comfortably. She rolled over, fitting her supple form into the bumps and hollows of her granite bed. At last she was forced fully awake by the gnawing emptiness inside her. She sat up into the breeze and sighed. Sonara wasn't only hungry for food; she was hungry for companionship too. For the second night and day she had been unable to find her sisters. She had become separated from them when she had chased and eaten a hare by herself; the others had gone on, hunting larger prey. Sonara had followed their clear scent trail for a long way but it had suddenly vanished, tracks scattering in all directions. Last night she had wandered far, finding neither company nor food. Daylight had found her near this rocky outcrop, lonely, tired and very hungry. She had climbed up for rest and shade, and to scan the countryside, as she was doing now.

The sun was low over the far hills that border the western edge of the great Serengeti plain. The restless day had brought no rain, only the hot wind that sucked the moisture from sky and earth. It was late in the dry season of the year and the ocean of grass stretched away empty, a rich golden-brown like Sonara's back as she sat scanning the landscape for food, friend or foe. She had survived three dry seasons so far, and knew them as times of hardship, of long dusty trudges between scanty meals and brackish waterholes. As she gazed over the evening landscape she saw little sign of life. A few gazelles grazed in the distance and a straggle of giraffes wandered among the thorn trees in a far-off dry watercourse.

A whirring sound made Sonara turn her head to look at an angular object

nosing near her clump of rocks. The car full of tourists passed by without seeing her and rushed homeward into the setting sun, trailing a brilliant plume of dust. She took little notice of it; cars were a familiar part of her world, noisy, smelly and inedible. She returned her gaze to the sweep of grass, rippling and whispering in the evening wind. From where she sat, she could see most of her homeland. To the south, hidden by the gentle rise and swell of the plains, the Seronera river had its source in a big marsh. From there all the way along the shallow watercourse as it snaked north to the woodlands was the best hunting during the dry season, for prey could be ambushed at the few remaining waterholes. But she dared not visit that stretch of dry river because hostile lions lurked there. Closer but almost as dangerous was a line of rock islands, the Masai kopjes. Sonara and her companions had been born among those sheltering rocks, three and a half years ago. It was there that she would go to seek them now.

The last embers of the sun glowed crimson and the cool breeze caressed and invigorated Sonara. She yawned, sunlight gleaming on her clean white teeth. Her rounded ears were cocked to catch any sounds; her amber eyes alert for the slightest movement. She stretched, curling her tail over her back and tensing her strong young muscles. Her sheathed claws peeped through the creamy fur of her round, padded paws, then hid again as she stood erect, relaxed and ready to go.

Quietly Sonara slipped down her rock, sure-footed among the smooth boulders. A thin scream slashed the silence and a dozen brown creatures the size of big guinea pigs scampered to safety in various rock crevices. She paused to poke her head into a crack and could just see the forms of the frightened hyraxes huddling there, well out of her reach. She went down, over the rocks and through the prickly bushes that grew on the slope of the kopje. When she was gone the hyraxes came out again to bask a little longer on the sun-warmed boulders. Soon they would have to retire to their nooks to hide from the terrors which darkness might bring, such as the old female leopard who so often came to these rocks. Like small furry rocks themselves, the hyraxes watched Sonara leap into the sea of grass.

With an easy, graceful stride, Sonara set off towards the distant Masai kopjes; it would be dark before she reached them. She knew this area very well and the coming dark would not deter her from seeking out her companions. But other lions also awoke at dusk, lions whom she would rather avoid. Alone, she felt anxious, and as she hurried along, she uttered low groans as a mother does when calling her cubs. She dared not make the full roaring sound that would tell her comrades where she was. Others not so friendly would also hear.

Not far from the kopje was a flat-topped acacia tree. Sonara headed for it and as she drew near a hyrax froze high in the canopy. Venturing away from the safety of the rocks he had been munching at the fresh, fast growing shoots that sprouted on this tree just before the onset of the rains. Food was scarce and the hyrax had taken a big risk to go so far and to stay out so late. But Sonara ignored him completely; she was too busy sniffing around the base of the tree.

Bush hyrax

Sonara sniffed the ground and grass and trunk. There was a vague scent of lion. Two adult males had rested there the day before. She did not recognize their scent, but as it was not fresh it did not particularly worry her. She reached up on the trunk of the old tree with her forepaws, as if to stroke it, raking her claws down the rough brown bark. The hyrax watched suspiciously from above.

When Sonara was much younger she had clambered clumsily on a branch of this tree but it was gone now, as was her childhood. She was fully adult, sharpening her claws and leaving her mark on the sturdy old tree. This particular tree was seldom visited either by lions or by hyraxes. It was just a little too far from the kopje. Lions usually preferred the solid shade of the rocks, and the wind and view from the kopje top, to the prickly dappled shade beneath the thorn tree. Hyraxes came rarely for it was dangerous to travel so far into the

Leopard

24

open grassland. Sonara dropped to all four paws and left the tree. So did the hyrax. He scrambled hastily down the trunk and hurried homeward through the forest of grass. But he never arrived; he did not see the old female leopard until it was too late.

Sonara glided through the dusk, her paws thudding softly on the paths hidden beneath the grass. She paced steadily in the long-distance gait that lions use when they travel purposefully but are not hunting; ploughed through the tall grass of a watercourse and bounded lightly over the dry gully. She continued on up the rise, fully intent on the distant kopjes. Suddenly she stopped and turned towards a small bush. A strong scent wafted from it. She sniffed at the bush and at the ground near it, wrinkling her nose as she inhaled the powerful odour of an adult male lion. The strength of the scent meant that he had passed here this morning, spraying the bush with his urine to advertise his visit.

Sonara moved on quickly. Now she was even more apprehensive. She recognised the smell of the male; he was one of five whom she feared. For many months, Sonara and her companions had been avoiding these five fierce Loliondo males. To her they were burly, bushy-maned brutes who chased her and her sisters on sight. So far, the young females had always managed to escape; they did not want to find out what would happen if the males ever caught up with them. Sonara was eager to get out of the area and rejoin the eight females who were her only friends in the world. Being alone and in danger made her especially nervous and she increased her pace to a swinging trot.

She jogged along until she grew tired, then resumed her long-distance stride. In the gloaming the Masai kopjes looked like bones sticking out of the lean landscape; Sonara was but a speck moving over the hide of the plains. She jumped in alarm as two dust-coloured birds fluttered up from her path on boldly striped wings. Shrieking, scolding, screeching, the crowned plovers rose on the

Crowned plovers

wind and swooped back low over her head, self-appointed sentinels of the grasslands. Sonara began to jog again, leaving the noisy plovers behind. She still had a long way to go.

Before the coming of the five males, Sonara had lived with her extended family, the Masai pride. As in all lion prides, the core of the Masai pride was a group of related females – mothers, daughters, sisters, cousins, aunts. At the time when

25

Sonara had eight "mothers" and twenty-three "sisters and brothers"

Sonara had been born, there had been eleven adult females, eight of whom had given birth to a total of twenty-four cubs. The cubs had all been born within the space of two months, during the rainy season, when there were many prey animals grazing the new grass on the plains. The females of the Masai pride were good hunters and had provided well for the mass of little cubs and for the two old males who had fathered them.

Sonara's mother had led her three cubs from their den in the Masai kopjes as soon as they were old enough to walk. That first journey from the rocky den to the Seronera river to join the pride had seemed endless to Sonara, who had mewed in protest for most of the way. The trickle of tiny cubs emerging from the kopjes became a flood as the eight mothers had pooled their litters. Sonara had soon learned that she had eight "mothers" to suckle, guard and play with her, and twenty-three "sisters and brothers".

Sonara was one of the lucky, robust sixteen cubs who had survived their first dry season. By then she had been weaned to a diet of pure meat. To feed themselves, the males and all the cubs with their growing appetites, the Masai females had had to hunt often. The cubs were frequently left behind until one or two females came back to lead them to the kill. While the adults were away and during the long periods when their parents slept, Sonara had come to know the other cubs better than she knew any of the older lions. The large set of cubs of the same age formed a close, companionable unit.

The cubs grew fast on their rich diet. As they became stronger and more active, they had learned the landmarks and the limits of the Masai pride's territory. To the north and west, the frontiers were marked by other large prides which their parents avoided or chased. To the east a small pride held the little cluster of Boma kopjes and surrounding plains. To the south-east the plain stretched open and free. In the wet season it was a green expanse luring the great numbers of animals from the woodlands until the plain was covered from horizon to horizon. The frontier out there was drawn by the seasons. When the grass was green and prey was plentiful, the lions might travel south-east along the Seronera river to the big marsh at the limit of their usual dry season range. Past the marsh there was a shallow valley that collected the water which became the Seronera river when it rained. There the Masai pride females might take their playful brood, roaming the open spaces, feasting and growing fat on wildebeest and zebra. But when the rains ended the herds left, the grass turned brown, the waterholes dried up and the lions returned to their normal range.

By the end of their second year, Sonara and the other cubs were taking part in hunts, learning the skills they must master in order to support themselves. Their days with the pride were numbered. Their fathers had disappeared and the five young Loliondo males had moved into the area. The cubless females of the Masai pride were the first to accept the males and mate. The mothers with cubs avoided the new males at first. However, after two years of rearing the numerous young, some of the mothers were ready to mate again. One by one, they deserted their large cubs.

Sonara and her companions tried to rejoin their mothers from time to time and were occasionally allowed to join in at kills, if none of the new males was present. The big males had no use for any of the youngsters. They were especially intolerant of the male cubs, who now had crests of hair sprouting from their heads and soft curly manes. These little males were attacked so often that they had to leave the area. Sonara never saw them again.

During their first year, the nine remaining daughters continued to live on the margins of their mothers' range. The young females were no longer allowed to share the Seronera river and adjacent hunting areas. All the old females now had new batches of cubs, sired by the five new males. Sonara and her "sisters" were definitely unwelcome. They had become wary, using all their senses to avoid being chased and attacked by their mothers or their mothers' new consorts.

Right now, Sonara was being especially careful. She was nearing the Masai kopjes, a long trespass into the heart of the pride's range. She moved carefully through the dry grass, straining to hear any sounds, sniffing the air and the ground, peering into the darkness. For some distance there had been no sign of any lion and she walked with caution. A trace of scent came to her on the cool breeze. Sonara paused and raised her head, inhaling deeply. She exhaled in delicious relief; it was not the scent of an enemy but the perfume of one of her sisters.

She began to trot, jogging through the grass and jumping a ditch beside an eroded tourist road. She ran up the wide sandy track towards the Masai kopjes. A movement out in the dark grass caught her eye and she stopped still. A pair of ears rose among the grass clumps. Sonara stood, uncertain. She sniffed the air but could catch no scent. She uttered a deep soft call, a low questioning moan. The owner of the ears sat up, then stood. It was a slim young female like herself; it was Swala.

Sonara galloped off the track and bounded through the grass. Other heads and bodies popped up all around. Ecstatically, Sonara rushed at Swala, crashing into her shoulder, knocking her over, falling on top of her. Swala grunted and slid out from underneath, then flopped on top of Sonara. Others joined in the greeting. They slid against her smooth sides, nuzzled her face, sniffed her and swatted her gently with velvet paws. They were all there: stately Swala, pale Shiba with the protruding lip, diminutive Safi who lacked a tuft to her tail, honey coloured Sega with the scarred face, sugar-frosted Sukari and playful, lithe Sarabi. As usual, Siku was half asleep in the grass and diffident Salama with the fuzzy ears stood apart from the rest. Salama waited until the others had calmed down a bit, then walked over and rubbed her face thoroughly against Sonara's. Siku finally woke up and came to slide against Sonara, draping her tail over Sonara's back and flopping beside her as she settled into the grass to lick the top of Swala's head.

The joyous reunion over, the nine females lay about in the grass, licking one another or looking around to decide where next to go. They were soon told where NOT to go. From the direction of the big marsh to the south came deep groans. They were repeated. The duet of roars from two of the Loliondo males bellowed out through the night. They rose in a crescendo, then shortened to a series of soft fast grunts made in perfect unison. For a moment there was nothing but the soft, ceaseless throbbing of crickets. Then another male began to roar from the direction of the river on the other side of the Masai kopjes. He was the only Loliondo male with the Masai pride females this night and answered his two brothers in the south, pealing his great booms of greeting thudding through the darkness. Some females of the Masai pride added their clear rolling tones to his, lazily accompanying one another. The tremendous sound filled the night as still more adult females joined in. By this time most of the young females were already on their feet, listening with full attention. Their tails flicked in agitation as the thrumming waves of sound from the chorus rolled from the dry river across the plains and echoed from the nearby kopjes. Gradually the choir hushed as one by one the elders of the Masai pride breathed their final flourishes.

The nine young females turned away from the river and the roarers. The night that Sonara had lost them, the other eight had narrowly escaped a severe

Sonara was greeted by her sisters

attack. They had left Sonara munching her hare and gone on hunting unsuccessfully until they had heard the Masai pride females growling over a fresh kill. Their hunger had made them dare to go close but the older females had chased them away at once, needing the food for themselves and their dependent cubs. Some of the Loliondo males had appeared during the commotion and chased the young females too. The eight had fled, scattering in fright. It had taken them two nights to come together again and Sonara was the last one to rejoin. The band of females turned north-east.

Almost directly ahead a pair of noisy crowned plovers scolded someone in the darkness. Then came the gruff grumble of the male lion that had disturbed them. The females stopped in their tracks as the smell of the male came to them along with the rumble of his first roars. He was far too close but luckily upwind of the silent females. Away they ran. It was Leo, the male who had left his mark on the bush that Sonara had passed in the dusk, and she took the lead in her haste to get far away from him. They ran quickly until the roars from Leo grew faint, then they slowed to a trot as his last grunts resounded across the grass. He was not following them; perhaps he had not even noticed their presence. But they kept moving. Blocked on all other sides, the nine young females headed south-east, out towards the open plains and tomorrow's rising sun.

Laibon

Lerai

Lemuta

Leo

Lengai

Nafasi

Kali

Kesho

CHAPTER TWO

Brothers and Loners
(Late dry season, 1st Year)

Leo finished his roar with a final grunt at the ground, his nose almost touching the dry earth beneath his outstretched paws. His paws, covered with sticky, dried blood, smelled tacky, accusing him of neglect. He began to clean them, first one paw, then the other. While he cleaned himself he waited to hear any response to his call. Roaring was the usual means of communicating one's location; the loud, low-pitched sounds travelled far on the still night air. Leo had broadcast his whereabouts to all listeners – he was near the Masai kopjes in the heart of the Loliondo males' shared territory. He listened as two of his brothers, Lerai and Lengai, began to roar. They were still far away, in the direction of the big marsh at the southern limit of their territory, but their voices were closer this time, they were returning to the centre of the range. Leo listened to their roars, continuing to clean himself.

He finished licking his paws clean, then used them to wash his bloody face, sweeping them over his cheeks and muzzle, listening to the final grunts of Lerai and Lengai. He did not answer his two brothers; they knew where he was and he did not intend to shift position for a while. He was feeling lazy after his recent meal, his belly full of the gazelle meat he had consumed. The gazelle had been stolen from several hyenas who had run smartly when Leo had rumbled up out of the dark. He had been wending his way homeward, re-marking spots along the eastern border when he'd heard hyena growls and suppressed giggles. The occasional confiscated gazelle or other kill was the price the hyenas had to pay for hunting rights within the lions' domain.

One last swipe at his face and Leo began to groom the top of his head. Using his big paws, he wiped his forelock over his broad forehead, cleaning the black

backs of his ears in the process. He paused in his grooming to listen to the roars of Lemuta, fourth of the five brothers. Lemuta's loud booms came from the direction of the Seronera river on the other side of the kopjes. He was obviously on the move, leaving the Masai females and cubs who lay scattered, resting near a pool in the mostly dry riverbed. Lemuta was on his way to join his brothers at the kopjes. Leo resumed his methodical cleansing. He would wait for his comrades to come closer to him before making the effort to move.

Leo started on the job of tidying his mane. He twisted his thick neck to the side and used his long raspy tongue to smooth the tangled hairs covering his shoulders. His woolly mane took a long time to groom, but properly combed it enlarged and enhanced his appearance, intimidating smaller, younger, or less fit males. His mane also protected his neck in fights, and fights were inevitable for Leo and his brothers. The five males never fought seriously among themselves, their aggression was directed towards strangers. Together, the five Loliondo males managed to hold a huge territory which they had laid out themselves and now held three different prides of females. Wandering males and neighbouring territory holders constantly penetrated this huge range. Intruders had to be taught severe lessons and if they did not retreat quickly enough they would be attacked. Usually the sight of one or more of the bushy-maned Loliondo males was enough to scare outsiders away, but occasionally those great shaggy mantles served duty as padding against ripping claws and tearing teeth.

Leo continued to ensure that his dual-purpose cloak was in prime condition. He could only just reach the front of his mane by arching his neck and sweeping his tongue sideways across his hairy bib. His chest hair was paler than the rest of his mane which was a mixture of muted colours, mostly buff and brown with

34

black strands on top and blond around his face. The colours blended well with those of the dry grasses around him but unless his head was well down, Leo's mane was no camouflage. A big mane could be a real hindrance in hunting, especially on the plains where there was little cover – prey could easily spot such a moving bush. Thus Leo and his brothers usually had to rely on opportunistic hunting or scavenging. Regular meals ensured that manes grew luxuriant, while meagre or infrequent ones meant that they grew tatty and thin; direct signs of the prowess of males. Leo's mane was beautifully thick and he looked neat and well brushed as he finished grooming (if you didn't look too closely at the cheek hairs still stuck together with blood).

A family of jackals barked at something and Leo turned as he heard low roars follow the yapping of the jackals. Lerai and Lengai were definitely much closer now, just south of the row of Masai kopjes. Their roars were answered by Lemuta who was also nearing the kopjes. Laibon, the fifth and last of the team, roared too, hurrying on his way from the Seronera pride in the north. Leo joined the round of roars. He stretched out his neck, grunted hugely like some engine stoking up, then made another louder grumble to get going. Inhaling deeply he bellowed out a full volume roar; six great heaves of sound followed in

succession, then the long descending series of short fast grunts. He terminated his performance with his characteristic long shuddering sigh and a final grunt at the ground, nose between his huge clean paws.

Leo stood up and looked over to Masai kopjes, black blocks against the starlit sky. Before he left to join his brothers there, he marked himself and the spot where he stood. His backward pointing penis ejected several squirts of scented urine directed at the ground. While squirting, Leo scraped the earth and grass with his back paws, getting some urine on his feet and on the underside of his tail. This procedure scented his tail, feet and the ground with Leo's personal mark and as he left the spot he trailed a cloud of odour readily recognisable by other lions. All adult lions, male or female, peed and scraped in this way, usually several times a day. It was a way of communicating which was more intimate than the widely broadcast roarings.

Leo set out slowly in the direction of the nearby kopjes. His full belly swayed from side to side, brushing the tops of the tufted grasses as he walked along. He reached the track leading to the kopjes and turned onto it, huge pads plopping along in puffs of dust. It would be good to meet with all his brothers again. The five males were seldom together for there were many duties to perform – checking, marking, patrolling, defending the range, mating with receptive females, eating and resting with the various groups of females and cubs. The five brothers shared these responsibilities, but their cooperation yielded a long-term reward – the cubs they sired. The more prides the males could hold, the more cubs they could produce, leading the brothers further afield, and more often apart. Leo increased his pace as he anticipated the reunion.

36

The Loliondo males were seven years old, in their prime. They were age-mates, born at about the same time to females of the Loliondo pride to the north-east. There had been six of them when they left their home pride at the age of three – callow youths with small manes. This unusually large group of young males had simply moved next door and taken over the neighbouring Seronera pride, ousting two old resident males. There the inexperienced Loliondo males had matured, taking on the responsibilities of pride males and siring many cubs. They had lost one member of their team, but remained a powerful unit, gaining an extensive knowledge of the area and learning how to defend it. After a couple of years they began to expand their range. They found the cubless and receptive females of the Masai pride who readily accepted the vigorous new males. The other Masai pride females, with large cubs, had run away at first but gradually the cubs had been chased off and the females won over. The Loliondo males had mated with the Masai females and already there were new cubs.

Extending their interests further, the males had recently found yet another pride in need of their attentions. The small Boma pride to the east was bereft of adult males, so the Loliondo males were beginning to make regular excursions eastwards. Leo was returning from a visit to the Boma females. While he and his brothers moved from pride to pride, each group of females occupied its own separate range. Female lions defended their ownership of a familiar area that produced prey while male lions defended their ownership of prides which yielded cubs and food. That was the structure of lion society.

Leo stopped walking. He heard a soft "humph" off in the darkness and turned to see two dark lumps moving past a pile of rocks. He waited. The shadowy forms of Lerai and Lengai grew larger. Leo grunted and was answered; he approached them sedately, head held high. Lengai greeted him first, sliding alongside, leaning his whole body on to Leo, then flicking his tail into Leo's face. Lerai simply nuzzled Leo's head and mane, then lay down to rest. He could smell the remnants of blood on Leo's cheeks and it made him feel hungry. He and Lengai had come a long way already that night. They had been on an excursion to the big marsh at the southern edge of their range to check for intruders. It had been a long trek and they were tired, and hungry. Sometimes patrolling paid off in food, but more often it only had the intangible reward of maintaining invisible boundaries. Leo and Lengai lay down near Lerai, all three listening to roars from the approaching Lemuta. Leo grunted as he sighted Lemuta coming between two kopjes; he rose to meet him. Lerai and Lengai followed tiredly. They met the newcomer with friendly face rubs and slidings. Leo had enough extra energy to be playful, collapsing against Lemuta and rolling over in a floppy embrace on the ground. The four males sat upright to listen to calls from Laibon who eventually appeared. More prolonged greetings followed because it had been a long time since he had last joined his brothers. Laibon had an odd mixture of scents on him by the time he had finished rubbing and being rubbed by the bodies of his friends. Leo sniffed him all over and rubbed his face against Laibon's once more, making sure the glands above his eyes marked the prodigal male.

The five reunited brothers moved off to the closest kopje. Leo headed straight towards a little bush at the base of a huge pile of boulders, put his head into it and sniffed. There was no odour of strangers, just that of his comrades. Any of them might stop to mark this particular bush when they passed. He lifted his head, brushing the leaves of the bush against glands on his face. Then he turned and backed into the bush, arching his tail up over his back. He sprayed the bush with urine, squirting it in a fine, dense mist all over the leaves and stems, giving his tail another touch too. With a few flicks of his tail he left the well-marked bush and went up the sloping rock of the kopje.

The strong smell of hyraxes hidden in the rocks hung like a blanket around the kopje, for they marked their own small property too. Leo went all the way to the top and stood majestically on the exposed granite cap. The wind ruffled his mane as he looked out over their kingdom. None of the others joined him on the exalted throne so Leo descended, satisfied that their joint territory was intact. His confidence didn't come from the limited view he could get from the kopje top in the darkness. During the time that he had spent in the east he had not smelled any other males. Nor had Lengai and Lerai found any sign of trespassers when they had detoured south to check that boundary. Lemuta knew that only he had been near the Masai pride and that no intruders had dared come near their

stretch of the dry river. Laibon knew that no strangers had crossed the northern perimeter of the dominion either. Their kingdom was entire.

An old moon peeked over the eastern horizon to glow over the plains. The five strong, proud males stood or lay around, looking like a collection of beautifully sculpted statues. Leo walked over and flopped on to Lengai, who squeezed out from underneath with a grunt. He repeated the grunt. The grunt became a rumble and finally a roar. Leo joined in, then Lerai. Lemuta and Laibon added resounding blasts and the impressive round of roars thundered out over the pale tawny land.

The roaring rolled across the plains in all directions, reaching a distant kopje to the north-east where two silent males perched. Kesho and Kali had remained quiet through all the earlier interchanges, listening carefully to the news. The messages had told them that their neighbours and enemies, the Loliondo males, were out patrolling or with their various females. Then all five had come together in the heart of their range, roaring together with strength and solidarity.

Kesho and Kali waited patiently for the Loliondo males to finish their proclamations of power. The two brothers felt safe on their rock, well away from their five foes in the west. Tonight they were on patrol through their range, a territory they had held for over a year now but were soon to abandon. Their pride was no longer very attractive to them. All the females had cubs and Kesho and Kali had grown restless. They were ready to move on. Recently they had found the scent of several young females in the neighbouring area and would follow up this new lead.

The Loliondo males were now silent, having finished their challenging chorus. Kesho led off a reply and Kali joined before the first bellow had left his brother. The two males roared slowly and sonorously, taking their time to produce each blast. The kopje rocks behind them caught and amplified the sounds, echoing them into the soft night. The defiant duet rang out grandly, gradually the final series of grunts petered out as the two brothers reached the end of their retort. There was no reply. Neither their own females nor their foes bothered to answer. But they did not mind, they had each other, their wits and five years of shared experience. They were energetic and youthful, just entering their prime years. Already they had spent a good part of their lives as wanderers, learning to survive as nomads, waiting for the chance to win a pride. They had not only survived, and won a pride, but had acquired skills in the arts of defence and wise retreat, of finding and confiscating food. They were a crafty pair and very close, having no other friends but one another.

They descended the rocks, marked the bushes at the bottom and set off across a long stretch of open grassland towards the lone kopje. Near the kopje was an old flat-topped acacia tree where they had rested briefly the preceding day. While in that area they had smelled traces of some of the interesting young females; they would return to check the spot again. Side by side, Kesho and Kali strode along, walking fast, their goal in sight.

40

Side by side, Kesho and Kali strode along

The roars of Kesho and Kali still echoed in Nafasi's head. The young lion lay petrified, close to the kopje where the two males had roared. His muscles were strained from his long, motionless crouch. He peeked over the stems of grass to risk a look around. One glance at the two shapes and he flattened himself into the grass. He waited tensely for the two big males to go away. Blundering confusedly through unknown country, booby-trapped with resident lions, Nafasi had been so careful, avoiding any lion in sight, staying away from any area where they roared, moving away from any fresh scent. Now here he was, trappped in a thicket only a few paces away from the dreadful sound and sight of two mature males. He dared another glimpse and saw that the males were moving away, shoulder to shoulder like two chariot lions, pulling their long moon shadows behind them.

Nafasi relaxed a little and nuzzled his paw. It hurt. He'd stepped on a thorn in his undignified flight from a group not far away. They had attacked him in the dark, snarling and leaping at him, scratching his rump. Terrified, he had run as fast as his young legs would carry him, crashing through bushes, across a stony riverbed, around a low hill. The scratches on his rear smarted but the thorn embedded deep in his paw hurt horribly.

Nafasi was being squeezed steadily out of the area. Only one place remained open for him, a place he had not yet visited, the vast plains to the south-east. He was one of those unfortunate male lions who had been forced out of his home pride with no brothers to accompany him. Abruptly, he was on his own, trying

41

desperately to feed himself and to stay alive in a land full of hostile resident lions. Luckily he was young, a healthy three-year-old who could run fast and with the brains to hide when necessary.

A long time seemed to pass. Nafasi licked his paw tenderly. He poked his head up again, peering into the dim, moonlit landscape. The two lion shapes were dark on the horizon, much smaller now. He stood and limped off painfully. He was a nomad, a drifter with much to learn if he were to survive. He slunk away over the ridge to the south-east, alone. Nafasi headed to the open plains.

Kesho and Kali did not notice Nafasi. They strode on through the grassland, crossed a dry stream bed and moved up the slope to the lone kopje. Before reaching the rocks they found the scent trail that Sonara had left. They checked the kopje and tree, then began to follow Sonara's trail. But it went into the kingdom of their five foes, so the two males turned south, backtracking on the scent until they accidentally crossed a treasure trail.

The sky was just beginning to grow light when they found the wide, scented trail. Nine maiden females, all together, laying down a recent and fragrant track, easy to follow, irresistible. Kesho and Kali followed the summons eagerly, heading away from their enemies the Loliondo males and all their other neighbours and past relationships. The dawn wind blew gently in their faces, teasing their manes, as they faced south-east towards the free and open plains.

Nafasi licked his paw tenderly

CHAPTER THREE

Sametu Valley
(Early wet season, 2nd Year)

A fat drop of rain thunked on to Kesho's side. A few more bombarded Kali's head. Nearby, other drops splattered the outstretched forms of Sarabi and Siku. Scattering droplets caught Swala, Safi, Sega and Sukari who lay criss-crossed on one another in friendly disarray. Further away, the shower dampened the backs of Sonara and Shiba who were already sitting up, blinking into the wet dawn. Salama still lay fast asleep in the pattering rain. Slowly the storm and early light eased across the heavens.

The lions were scattered along an open ridge that ran from north to south, dividing the Seronera watershed on the west from the Sametu Valley on the east. Since the onset of the rains, three months earlier, the nine females and their two new male companions had been roaming the outer reaches of the broad Seronera valley. As the abundant prey gradually drifted past them en route to pastures in the south-east, the lions had wandered that way as well. Recently they had been making use of the long dividing ridge with its view into the two valleys and a small waterhole near a rocky high point. The ridge was a good temporary base, but the wanderers were soon to discover a far better home.

The dawn rainstorm steeped the stained sky and suddenly poured down its brew. All eleven lions sat up and hunched against the rain. They looked like sodden grass clumps on the wide Serengeti plain. With manes like great mops, Kesho and Kali seemed the most miserable as they all waited patiently for the storm to swirl away on the morning wind. Gloriously the sun peered over the far eastern rim of hills, the distant and ancient Gol mountains. It shone brightly on the clean land and made the hides of the lions steam. Kesho and Kali shook

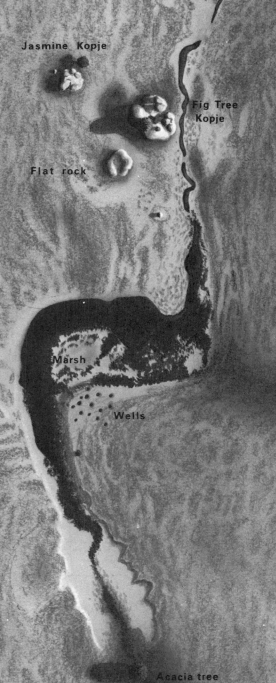

Map of

SAMETU

100 metres

Jasmine Kopje

Fig Tree Kopje

Flat rock

Owl Rocks

Marsh

Wells

Acacia tree

The kori bustard's display

out their manes. They shook the rain off their bodies and flicked water from their tail tufts. The nine females were doing the same. The lions started licking each other; heads, faces, necks, shoulders. Drops of water sparkled in the brilliant light. The nearby kori bustard untucked himself and shook the drops from his wings. Then he began to puff himself up, soft feathers drying quickly in the early breeze and warm sun. He preened and ruffled his feathers. At the edge of the close-cropped top of the ridge near the waterhole he spied his mate and he heard the distant booms of his rival. It was time to get to work proclaiming his territory. The big bustard folded his tail over his back, puffed out his white throat and paraded along the ridge. Near the rocky crest, he lowered his tail and standing very tall, began to boom; six deep rhythmic thuds as if to say ...'I'm ... a ... ko-ri ... bus-tard!'.

The lions and plains were taking longer to dry out but the strong morning sun was helping. A puddle near Sonara's shoulder was rapidly disappearing, the sun sucking up the moisture while the fine, porous soil sucked it down. By the time Sonara had been licked and dried by Sarabi and the sun, the rain pool had gone, evaporated, absorbed by the air, soil and the roots of the thirsty grasses. Sun, soil, water and grass – those are the essential ingredients that make the Serengeti plain such a special place. The soil on which Sonara and the others lay had come all the way from the Ngorongoro highlands in the south-east. Long ago, those blue mountains were fiery volcanoes. For more than a million years they had blasted ash and dust which was borne on the constant south-east wind and settled softly, slowly, year after year. The ancient granite plain was thus buried under a blanket of fertile soil, rich in minerals that all living things need.

The same south-east wind had brought this morning's storm and still ruffled bustard feathers and lion manes as it sped inland from the East African coast. The wind brings rainclouds carrying loads of moisture from the ocean

47

The migrants: Thomson's gazelle, wildebeest, eland, zebra

but most of the burden is dropped on the high mountains that border the great plain, the mountains that used to be volcanoes. Here and there on the plain the tops of older hills stick their rounded heads above the new soil. They are the granite kopjes upon which lions love to lie. These isolated rock piles capture an extra share of the rain so can support their leafy, temporal companions; water-loving grasses, bushes and a very few trees. The rain that falls on the open plain abundantly for only half the year is absorbed into the shallow soil and drains rapidly. Throughout the rest of the year, rain comes rarely and there are long

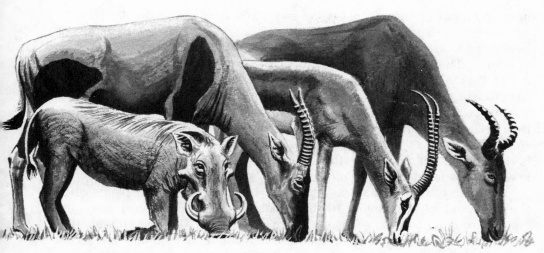

The residents: warthog, topi, Grant's gazelle, kongoni

dry spells. Only certain plants can cope with these conditions and the most successful are the grasses.

Serengeti grasses are special, like the soil in which they grow. There are many different kinds, each with a skill for survival. Some sleep through the dry season as fleshy roots, others as seeds, but all are awakened by the rains and begin a frantic race to cover as much ground as possible. Some can do this by growing thick and tall, choking their competitors; others send long runners in all directions and yet others spring up quickly from seeds and colonise bare patches. There are specialists which can cling to loose sand, or suck moisture from the soda-saturated soil beside waterholes. A few grasses gain an advantage by being unpalatable. One tastes terribly of turpentine, others have tough fibres or crystal teeth to lacerate grazing mouths.

There is even a grass that thrives on being eaten. Sonara and Sarabi were lying on such a grass now, a flat rug of green. For many thousands of years animals had come to the Serengeti and chewed a living off the grasses. The mat-grass had exploited this situation. When nibbled flat by grazers this kind of grass spreads out and grows quickly. The grazers clip down any other surrounding grasses as well so the spreading mat has room to grow. If for some reason the grazers don't come, the neighbouring grasses can grow tall and begin to shade and crowd out the flat grass and it loses its vigour. So the spreading grass actually needs to be eaten to thrive.

Sonara just knew that the grass was a most comfortable, soft and springy mattress on which to lie. She was more interested in the grazers that come to the Serengeti plains to eat the grasses when they are green and growing most vigorously. The grazers are remarkable too. Each has its own way of cropping the vegetation. The wildebeest or white-bearded gnu had a lawnmower mouth. Its wide, flat lips press the grass leaves into the mouth to be cut by the row of sharp lower teeth. Thousands upon thousands of gnu come to the Serengeti

49

plains during the rains and mow their way across the green sward, keeping the grass growing strongly, young and tender. The small shoots that they miss or cannot get at are nipped up by the neat little lips of the Thompson's gazelle or "tommy". The coarser grasses left alone by gnu or gazelle are chomped by zebras. Zebras have more teeth to cope with the roughage, but their digestion is less efficient than that of the cud-chewing antelopes, so they must eat greater quantities. This is why they always look so fat, their stripes bulging around their bellies. But zebras will eat soft, young grass when they can get it and so do elands, the huge yet graceful antelopes that migrate with the others. Elands eat herbs as well, which none of the others relish. Though rich and nutritious, herbs are scattered and very hard to find.

Wildebeests, tommies, zebras and elands have to abandon the plains when the grass dries up. They migrate to the woodland areas where there is grass and water throughout the dry season, which they share with the animals that reside there all year round, such as buffalo and impala. The small tommies are usually the last to leave the plains, reluctant to quit the short grass and flat, open spaces where they can see and run easily. Some other grazers also try to remain on the drying plains; just a few can be found here and there throughout the year, making a living off the old, tough grass, infrequent green shoots or herbs and anything else they can find. These are the residents of the plains and include topis and kongonis, Grant's gazelles and warthogs. But when it gets very, very dry, even these have to seek better pastures. The larger migrants and residents share the plains with smaller animals, many kinds of insects and birds, each having a place in the panorama of life on the Serengeti. Sun, soil, water and grass; then the many grazers large and small. And finally, those that eat the grazers, such as the bustards, searching for grasshoppers along the ridge or the lions lounging, not actively searching for anything, yet.

Kesho finished licking Kali's mane and stood up, stretching. The fresh young morning beamed over the sparkling expanse of jewel-green grass. Sarabi reclined near Sonara on her soft green mat, watching the males. For three months the two brothers had been tagging along but Sarabi was still vaguely uncomfortable when they were close. All her sisters had mated with the males repeatedly and seemed to accept them as members of the group. Inner promptings made Sarabi bolder; she stood, then cautiously approached Kesho. He turned towards her. Sarabi was a relative newcomer to the mating game and was uncertain how to make her interest most obvious. But Kesho was a well-practised lover and knew immediately her intent. He waited patiently.

Sarabi walked past Kesho, flicking her tail up in front of his face. She went on a little way, paused to pee on the wet earth, then walked on. Kesho lumbered after her, stopping to sniff at the spot which she had marked. He inhaled deeply, lifted his head high and drew the scent into his nose, then wrinkled his muzzle and opened his mouth in a strange grimace, as if to taste the delicious fragrance. Once again he lowered his shaggy head to the ground and took another whiff. There was no doubt that Sarabi was ready to mate.

To ensure he got the message, Sarabi came sidling back to Kesho. She dipped under his chin and slid against his hairy chest, then turned and hurried away up the slope towards the booming bustard, Kesho following. Sarabi stopped and lowered herself to her belly; her tail flicking back and forth, slapping the ground. Kesho came close and sniffed her tail, getting a slap in the face from her tuft. Sarabi stiffened as he touched her, rose and walked around Kesho in a half crouch. Kesho stood still. He had been through all this so many times before. Sarabi's tense, unsure behaviour was typical of sexually inexperienced females. This was only her second time as his partner, but already she was less frenetic in her solicitations. Kesho and Kali had both been very patient with the many maidens. They had let the females come to them, never intruding at first, just being there, following at a distance, then coming closer and closer. One by one the young females had become interested in their new companions and had behaved as Sarabi was now; hesitant yet eager.

Sarabi continued around Kesho in a kind of slinking dance. She pressed under his chin, then turned her rump to him, swishing her tail and lowering into a crouch. Kesho walked to her and nuzzled her rump at the root of her tail. Sarabi leapt up and bounded away with Kesho at her heels. Meanwhile, Kali was occupied with similar manoeuvres. His partner was Siku. The two couples gradually drifted away from the rest of the group as the two females led their consorts up the slope to the kori bustard's stamping ground. The bustard

strutted away slowly. He kept his distance, for although he could fly it took him a long time to lift off, and he liked to keep a good runway between himself and an approaching lion.

Those particular lions were far more interested in mating than in bustards. Sarabi went into a crouch again and this time could just bear to let Kesho cover her back with his body. But as he slowly lowered his muzzle to her neck and his haunches to her rump, she bolted again. Kesho plodded along behind. A little way ahead, Sarabi crouched; she held still as Kesho mounted and gently licked the back of her neck. He shifted his hindquarters so he could thrust against her; she accommodated him by moving her tail aside. At the moment that his penis entered her Sarabi growled low, a deep rumble; Kesho made a whining snarl, high and throaty. He held her loose neck skin lightly in his mouth, thrusted more deeply and quickly, and ejaculated. Immediately Sarabi rolled on to her back, and swatted Kesho in the face, jumped up and ran off. Kesho accepted the blow, then padded in pursuit to where Sarabi was rolling on her back in the grass. She growled at him as he approached. With a huge sigh, Kesho lay down next to her.

With patience, Sarabi would come to accept Kesho completely. Non-threatening, non-retaliating, gentle but always persistent, he would follow her. Kesho and Sarabi had much mating to do. Copulating several times an hour they would mate the day away, and tomorrow, and the next day too. Kali and Siku would do the same. Unlike the kori bustards, gazelles, zebras and so many other animals around them, the lions could afford to copulate again and again, for no one hunted them. A meal now and then of pure meat gave them the energy to spend whole days and nights mating. For young females with strange males, this extended mating was especially important, creating the bonds necessary for the complex organisation of a lion pride. Powerful and aggressive males and females both needed to co-operate or at least tolerate one another rather than chase or fight. What better way to get to know each other than consorting for days, if one could afford to? Lions could.

Down the ridge from the "lovers" the rest of the females lay at ease. All of them were already pregnant, tiny cubs reposing inside, well-protected and unnoticeable. Sonara lay with her head pillowed on her paws, sleepily gazing south along the ridge. She sat up abruptly when she saw two zebra stallions racing along the ridge, braying loudly. The one in front bucked and kicked at his pursuer, then galloped off. They both went over the ridge, across a wide saddle and on down the slope to the west. Sonara watched them until they

merged with the throng of shifting stripes belonging to other zebras. Many zebras moved along the slope, grazing the new and old grasses, still beaded with raindrops. Beyond them, the slope led in a steady, slow sweep down into the wide outer Seronera valley, the extent of Sonara's familiar range. Big as that valley was, and good for hunting, there was nowhere safe and secure to have cubs, no reliable watersource except the big marsh that was too close to the Masai pride of lions and the dreaded Loliondo males. With the vague stirrings of the cubs inside her, Sonara licked her flat belly and the now slightly protruding nipples.

She raised her head as a swift movement caught her eye. She stood and saw a vulture drop into the small Sametu valley just down the ridge to the east. Shiba had also noticed the vultures and rolled upright as another distant speck descended rapidly out of the clear, rain-washed sky, followed by yet another. The vultures were landing somewhere out of sight in the bottom of the valley. Sonara promptly set out at a trot. The possibility of a free meal was always appealing and vultures were a fairly reliable sign of free food. Out of the thousands of animals that came to the plains in the rainy season, a certain number died of old age and various ailments. The sharp eyes of the vultures were always open for such banquets. They could scan the ground from on high and were often the first to find the carcasses of the dead and dying.

Shiba gained on Sonara as she followed behind at a quick trot. The two trotted together down the long slope that led into the little Sametu valley. The vultures were landing near a small lake lying like a mirror in the valley bottom. Sonara and Shiba slowed to a walk. The sun was steaming up moisture from the wet earth and the distance was further than they had thought. Behind them on the ridge crest that they had left, three more heads sighted the vultures and Swala, Safi and Sega started to follow Sonara and Shiba down into the valley. Steadily the lions moved down the long incline into Sametu as more vultures descended.

The lone white-backed vulture glided along. Ahead she saw another of her kind drop. Food. Her keen eyes scanned the ground below, missing nothing. Lions lying on a ridge top. They looked dead but probably weren't; further on, more lions, some walking towards rocks and a glinting lake; near the lake, birds crowded around a zebra carcass. The vulture folded her wings slightly and plummeted down to join them. She banked a little to make a perfect upwind approach. Scaly feet down as the ground rushed up, she braked with her wings and tail at the last moment and bounced on to the grass at the edge of the scrimmage. Twenty or more or her kind were already flapping and heaving around the dead zebra. They gabbled and hissed and shrieked, while a tall marabou stork watched them disdainfully from a distance. Despite his huge beak, he had to wait until the vultures tore off bits that he could then steal. The white-backed vultures were usually the most numerous at any kill and certainly the noisiest. The newcomer, wings outspread, gave a hissing screech and prepared to wade into the fray. But she had to duck as a larger vulture

The lone white-backed vulture glided along

skimmed in from behind, cuffing her head with a wingtip and landing recklessly amidst the whitebacks. This handsome, mottled bird was a Ruppell's griffon vulture, whose greater size gave him assurance. He elbowed and pecked his way through the mob around the zebra and plunged his head into the gory hole beneath its tail.

The carcass lay in a bare spot where the earth had been eroded by many hooves, wind and rain. The dead zebra had come to the nearby pool in the valley bottom to drink and had moved off with his fellows when they left after a satisfying slosh through the water. The old stallion had gone only a short distance from the little lake when he had collapsed and soon died. For months he had been weakening inside since another stallion had kicked him hard in a fight over his mares. The old zebra was now being reclaimed into the system, bit by bit becoming part of those that fed on his remains. The vultures were the first to transform the dead into new life. The lions came next.

Sonara arrived first, charging the fluttering mass of vultures. The big birds hopped and flew in all directions. Shiba ran up close behind, and both females started to feed immediately, slicing the tight, striped belly skin with their side teeth. The many vultures had not been able to open up the carcass with their hooked beaks and the lions achieved that quickly and neatly with teeth designed for the job. Before long, Swala, Safi and Sega joined in. The carcass was half finished by the time Sukari and Salama appeared, having at last realised why all the other females had disappeared over the ridge. The two courting couples remained on the bustards' hill, unconcerned about food.

The meat was fresh and good and the seven females ate their fill. Some vultures stood around, waiting not so patiently for their share, hissing at one another and tidying their rumpled feathers. Others had given up and flown off, for food was plentiful and the day was still young. The sun climbed overhead and the satiated females gradually quitted the remains, almost uncomfortably full. Sonara finished cleaning her face with one last swipe of her forepaw and rose. It was getting hot and she was thirsty. She turned to the nearby pool for a drink. As she approached the edge of the lake across the small sandy shore, the resident pair of blacksmith plovers clinked repeatedly in alarm. An avocet made haste past a blackwinged stilt to the other side of the lake but a pair of Egyptian geese merely swam away quietly, watching Sonara warily as she bowed to the water to drink.

Shiba joined her and they lapped side by side. Sonara finished first, disliking the salty taste of the mineral water. She stood up, scanning the lake, seeking a comfortable place to rest for the day. On the opposite side, the lake was bordered by rushes, then a long sweep of grass led up to a cluster of small kopjes. Beyond one end of the lake stood two more kopjes and at the other end was a tall, shady acacia tree. Feathery leaves grew between long white thorns and its lowest branches brushed the coarse grass that grew in the valley bottom. It was a yellow-barked acacia or "fever tree" which normally prefers river banks; its presence meant that there was water and root room underground, two commodities which were rare on the plains. Sonara watched a vulture flapping into the crown of the tree. She rejected the hot enclosed bower with its stench of vulture droppings and started around the lake to explore the opposite side.

Shiba followed Sonara through the clumps of stiff, green rushes that formed a small marsh. They went up a bank and followed the curve of the lake through an arc of grass that hid some odd formations. Here and there were holes and hollows. Some held rain water, some caught the mineral-laden runoff from the upward slopes, some cut through the compacted crust lying under the soil and held clear, fresh, spring water. Sonara stopped at one of these wells to drink deeply. Shiba found the place where the underground water seeped out of a natural spring among rushes. The water that supplied the spring, the wells and the big acacia came from beneath a hard rock-like shelf that underlay much of the Serengeti plain. On top of the shelf was the shallow layer of soil in which the grasses grew, dependent on rainfall. Few living things could tap the deeper supply of water except where the crust was thin or broken as in this valley

Birds at the marsh: Egyptian geese (left), blacksmith plover, avocet, black-winged stilt

bottom. While the spring seeped from a natural break, the well where Sonara drank was purposely made by humans, long ago.

Many hundreds of years in the past some people had come to the plains with their cattle to share the rich pasture during the rains. Reluctant to leave when the plains dried up, the people had the ingenuity to prolong their stay for a while by hacking through the hard layer in the valley bottoms near natural springs. By making wells they could supply themselves and their livestock with fresh water for a time, until that source dried up too and they departed, only to return with the next rains. Centuries later when the originators had mysteriously disappeared, the Masai people had rediscovered and reused the old wells. They had named this particular oasis Sametu.

Sonara finished drinking and looked around over the small marsh, the wells and shallow dips, up to the kopjes. A curve of bare-looking rocks formed a crescent up-slope to the east. Too far to walk in the hot sun with a full belly. Closer, in the valley bottom beyond the other end of the lake and a wide bare space, were two clumps of rocks, one compact and bushy, the other larger and more open. The bigger kopje was just right – a giant smooth boulder faced by a steep tall rock overhung by a many-branched fig-tree. There Sonara was offered a view, a breezy lookout and maybe even some shade.

Sonara and Shiba headed for the fig-topped kopje, skirting the little marsh, following the trickle of water as it struggled from the lake through mud and grass, on down the valley. Sonara leapt gracefully across the sluggish stream and ascended the big kopje. Shiba paused at the base to sniff in the long grass. There was no scent of lions. She went up the sloping boulder to join Sonara, who had already explored the top. The two contented lions looked back over the curving lake to the others who were still grooming themselves near the remains of the zebra.

Their paws padded over a million years of history

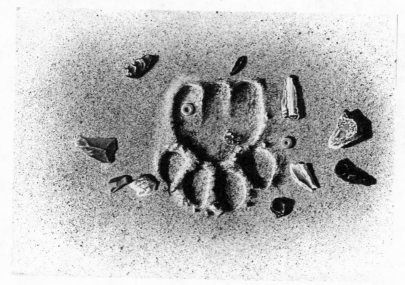

Sonara at the well

Swala looked over at her two sisters on the kopje, golden shapes lounging comfortably on top of the shiny grey rock. She went to join Sonara and Shiba for their midday rest, keeping to the open side of the lake. Safi, Sega, Sukari and Salama came along too, abandoning the carcass to the vultures. The lions crossed patches of grass and the many trails that carved their way down to the lake. They came to a long, flat, exposed rock close to the fig tree kopje. Between it and the bigger rock was a barren sandy place. Unknown to them, their soft paws padded over a million years of history. The small grey stones on the bare earth were fossil teeth and bones of ancient antelopes. Glittering black flakes of obsidian were left by the early people of the plains, perhaps those who had made the wells. They had brought the glassy rock from the distant volcanoes and chipped it into razor sharp blades for knives and spears. A few bright beads – red, blue and yellow – were more recent remnants of a sparse but decorated

Masai at Sametu

culture. Once the beads had formed a halo of colour around the necks of the elegant women who curved them into collars or sewed them on to the hides they wore as skirts. The Masai people had lived at Sametu for only a few generations but had left such signs of their residence as the wells, beads, bare spots and thick herb growth where their cattle had been kept near the kopjes.

Swala sniffed among the fragrant herbs fringing the skirt of the huge rock which the others began to climb. There were some old bones at the base of the rock, dragged there by porcupines, hyenas or humans long ago. She smelled nothing dangerous and turned to the kopje, lured by the smell of fresh water. In a cleft on the top was delicious water from the morning storm and Swala lapped contentedly, though her full stomach could hardly take in anything more. When she had had enough she went over and rubbed Sonara's sleepy face, then flopped on her sister, who grunted and moved to let Swala stretch out beside her along a seam of rock.

The seam along which Sonara and Swala lay held the roots of a row of stunted fig trees that had begun to grow after the Masai had left and had stopped using all available small wood for fires or fences. The baby figs had a good hold now with their curving roots deep into the long crack. Above them spread their mother, the big kopje fig which had managed to squeeze its many roots tightly into the junction of the two massive boulders that formed the bulk of the kopje. The fig tree's branches made a leafy umbrella over the steepest rock which none of the lions tried to climb. This was lucky for the pair of Egyptian geese who had their nest in the litter of sticks and leaves beneath the fig's long arms. The lions might well have eaten the treasure kept there, seven pale eggs laid in exquisitely soft goose down.

Egyptian goose nest

The females kept to the lower, more accessible platform of rock, lying in the open in the wind or under the paltry shade provided by a few small bushes or the row of baby fig trees. One by one the well fed, watered, pregnant females fell asleep. Sonara took one final look at the carcass and watched the last of the vultures depart. She gently pulled her foot out from under Swala's head and rolled over on to the smooth grey rock, streaked with pale rosy quartz. She and the others were soon fast asleep in the centre of their newfound home.

The big griffon vulture faced the wind and ran, beating his great wings, labouring clumsily into the air. Taking off was hard work. He flapped heavily towards several other vultures whose easy circling marked an invisible air thermal. Soon he felt a bump from below and warm air streaming up around him, lifting him high. He spiralled up, hanging weightless on outstretched wings, skillfully controlling his flight by subtle twists of his wingtips. The sparkling pool with the lion-draped rock at one end and the solitary acacia at the other receded as he rose higher. When the clouds loomed close overhead, the vulture left the thermal and headed south in a long glide. His crop had little room for any more food yet he automatically scanned the ground for carcasses. His skimming shadow overtook the waves of wildebeest lapping across the grassland, the small buff dots that were gazelles, and the light-grey specks that were zebras.

The wind sang through his wingtips as he flew over the vast herds. Beyond a little hill in the middle of the plains a circle of wildebeest framed an empty space. A single lion crouched in its centre, eating something. The vulture dropped a little for a closer look. It was only a tiny carcass, and not worth waiting for. He circled to gain height again and headed east into the wind. It was a long way home to the traditional vulture gorge where his mate brooded her single egg on the sheer pink and mica flecked cliffs of the ancient Gol mountains.

Down on the ground, Nafasi finished his meal. It had been but a tasty morsel, a newborn gazelle he had found crouching in a clump of coarse herbs, its mother far off. Nafasi cleaned his face, gazing at the watchful wildebeest who honked and snorted at their enemy. They and the other grazers knew exactly how close they could safely come and milled about restlessly, keeping Nafasi in view. The young lion moved off, opening a broad path through the wildebeest. He walked without a limp, because his paw had healed soon after he came to the plains three months ago. Those months had been good and the rich diet he was managing to get by hunting and scavenging had made his mane grow faster. In fact, Nafasi was becoming a very handsome male. His ears and face were unblemished and his body well proportioned, its burnished gold colour accented by his blond mane. He walked on with muscular grace until he found a flat mat of the spreading grass, and settled down on it to rest for the day.

Not far away was a little hill with some trees for shade but it was too far to walk in the hot sun and anyway two males had temporarily claimed the hill as their territory. Most of the lions that wandered the plains had no territory, finding food where they could when it was abundant and leaving the open spaces when their prey retreated to the woodlands. But temporary territories were commonly held by the more vigorous teams in the best places – spots with cover, water and a view as well as good hunting. Nearby Naabi hill was one such good place. Alone, Nafasi didn't qualify as a team although he had met several of the nomads with whom he shared the wide plains.

All the nomadic lions were outcasts from settled prides. They were a motley lot; scarred old males past their prime, furtive females with no kin, gangs of innocent two year olds barely able to hunt. There were also some fine young males like Nafasi, whose only major defect seemed to be that they were alone. He was learning, though, that not all strangers were hostile. A few females had solicited his attentions and recently he had met another young single male at a kill and they had fed together peacefully enough. In time such encounters might lead to friendship, and two males would be better able to take over a kill, a shady hill or even a pride.

Nafasi

CHAPTER FOUR

Cubs
(Mid to end wet season, 2nd Year)

Sonara plodded homeward, taking her time, stepping heavily on the dawn-dewed grass. Beside her walked Safi and behind her came others. Sonara's stomach bulged with the good meal of the night, pressing against the unborn cubs in her womb. The four milk-filled nipples that protruded from her belly like an udder would soon be put to use, as Sonara would give birth to her first litter of cubs that very night. She walked along slowly and Safi passed her by, her belly only slightly flatter than her sister's because she already had three baby lions tucked away in a den at Sametu.

Someone caught Sonara's rump and almost pulled her to the ground. She whirled around and cuffed Sarabi who plunged away to leap upon Siku. Sarabi's belly was a striking contrast to all the other females, being smooth, flat and with tiny nipples. Sarabi was the only one not to have conceived, and she retained her boisterous, playful humour while her sisters' attentions turned to the tasks of motherhood. Finding no responsive playmate, Sarabi rushed ahead, pounced on Safi in passing and reached Sametu before sunrise.

She waited until Sonara and Safi caught her up then the three continued along the valley to Fig Tree Kopje. They saw Swala lying on top and Sonara and Sarabi went up to greet her. Swala had been the first to give birth to cubs and had her two hidden below the guardian fig. She kept herself between them and her visitors but let Sonara and Sarabi sniff her all over. Swala smelled of milk and cubs, her own personal scent and the pride's special mixture. Swala rubbed each head then left Sonara and Sarabi to return to her cubs behind the boulders and twisting roots of the fig.

Sonara and Sarabi watched Swala disappear into her den then settled to rest on the cool rock. The sun had risen and was warming the day. Shiny black

wasps were beginning to buzz and rattle around flowering bushes and a lark sang out a lilting melody. Sonara laid her head on Sarabi's warm flank and gazed down from the kopje top at Siku who had decided not to climb up to join her friends. She crossed to the next kopje which was smaller, covered with wild jasmine vines and scored boldly by a split curving round its bulk from the ground to a clump of bushes at the top of the cleft. The bushes hid a lion-sized nest where Siku would make her den and she went there now to explore the site thoroughly.

Beyond Siku, up the slope, Sonara noticed Kesho and Kali coming home too. Other members of the pride had already returned, most heading to the wells and spring to drink. Safi, however, was hurrying to her den in the cluster of rocks to the east. As Safi hastened along, two pairs of ears tracked her progress. The resident Sametu steinboks sank immobile into the long grass as Safi passed them. The lions seldomed noticed the pair of delicate antelopes and did not hunt them, being more intent on the shelter and water provided by Sametu. But the steinboks always noticed the lions and took great care to avoid attracting attention. Safi did not see them and continued on her way to the crescent of small kopjes.

She approached the middle cluster of boulders across a lawn kept close cropped by the colony of spring hares who were hidden for the day down their burrows dug deep into the soil. An isolated stone in the middle of the lawn was capped by a wheatear, merrily singing his rights to his own small territory where his nest was hidden in a disused burrow. Safi passed him, avoiding holes and unseen inhabitants, beginning to sigh out a deep long call to her cubs.

Two spotted eagle owls swivelled their heads as the lion came near their small tree tucked in the lee of the kopje. The spindly tree was one of few protected from the wind in these wide, bare regions, and they were reluctant to leave it. Their great yellow eyes watched Safi warily as she reached their resting spot. When she approached the tree they flew off silently to land on an exposed rock. Safi went into the kopje, calling, and joined her cubs hidden in a shallow cave under an overhanging boulder.

Capped wheatear

68

Spotted eagle owl

The owls did not like flying nor perching in the sun and shifted from one taloned foot to another. A passing kestrel spotted them and buzzed them several times, causing them to duck. The owls flew off to their second best resting place, to the bushy back of the jasmine cloaked kopje where Siku was lying on the stony crest. They circled the kopje eyeing Siku and had to reject the hope of roosting in the bushes near her. Reluctantly they flew back across the valley to their original perch where at least Safi and her cubs were hidden.

The sun was heating the grey granite on which Sonara and Sarabi lay and Sonara shifted to try to find a more comfortable position, headfirst into the breeze. But she couldn't rest easily with her doubly full belly so finally sat up, leaned against Sarabi and looked thirstily at the lake. Heaving to her feet she stepped carefully down the sloping boulder, paused at the bottom to peer into a buzzing blue flower and plodded on across the bare earth, hot now in the midmorning sunshine.

The gabble and honking of the resident pair of Egyptian geese caused Sonara but a moment's hesitation as they led their seven fluffy goslings at a rapid pace across her path and into the water. They paddled madly away from the unconcerned lion. Sonara kept on, going around the lake to the area of the wells where she stopped to take a long drink. Kesho and Kali were lying nearby, having both tried unsuccessfully to fit themselves under a pair of stunted, spiky branched trees beyond the wells. Sonara went to greet recumbent Kesho who stood up courteously. She sniffed him then turned, swishing her tail, wafting him her scent. She also flicked her tail in Kali's direction as she passed by and went into the marsh.

Egyptian geese

Sonara chose a large clump of rushes in the middle of the marsh and climbed onto it, squashing the stiff stems flat. The sandy soil between the clumps was perfectly dry but the rushes could still feed on the underground water, remaining lush and green, making a stout bed for the resting lion. Sonara slept through the windy afternoon and did not join the others who set out to hunt in the evening. She gave birth to her cubs on a mattress of rushes surrounded by a thick screen of clumps, serenaded by roars from Kesho and Kali who were calling to the hunters from the other side of the lake.

Her first cub departed its warm womb in an envelope of thin membrane, eyes tightly shut. Sonara licked the cub free and dry as it squirmed on the smooth bed of flattened rushes. She helped another encased cub into the world, then another. Each cub received the caress of Sonara's tongue as she cleaned them thoroughly. Sometime later, the cubs nestled into her taut belly and found the nipples bursting with rich milk. Sonara stretched out among the clumps, feeling relaxed and peaceful as she felt their little mouths suck. The newborn cubs soon filled their small stomachs and curled up to sleep. Sonara looked at the spotty balls of fur in the quiet starlight, licked one's rump, then lay back and dozed with them.

Sonara stayed with her cubs all the following day, virtually invisible in the marsh, seen only when she climbed on top of the rush clump to catch the breeze in the afternoon. When she moved away from them the cubs instinctively crawled under any adjacent cover, creeping into the very heart of the rushes so

they were completely hidden. In the evening, Sonara went to the wells for a drink and to join the others briefly before they set out hunting. She stayed at her den through the night but left at dawn to run to join at a nearby kill, bolting some meat, then returning to the cubs once more. One cub was sluggish and did not find a place at her nipples as fast as the others. Before the first week of their new life passed, that weak cub languished in the unreachable interior of the rush clump and died.

Sonara moved the remaining two cubs to another clump, picking them up carefully in her mouth. At one week of age the cubs had opened their eyes to look short-sightedly at their world. To the swaying, dangling cub held in Sonara's mouth as she climbed over the rushes, the world must have looked peculiar indeed. She laid the cubs gently in another nest near Shiba who had also given birth to a litter in the marsh. The two cubs settled in happily, snuffling around their new home. They had soft blue-grey eyes, quite unlike the hard amber of their mother's who lay beside them. Their short fur was spotty and dense, their tails small and vaguely striped.

The cubs were very uncoordinated at first and their toothless mouths were easy on Sonara's nipples. By the second week they responded to her calls when she returned to the den to suckle them. With yelps and miaows they would

crawl forth from the base of the clump and greet Sonara with sliding and nuzzling, receiving licks and sniffs in return before they settled down to suck. By the third week the cubs' tiny incisors could be felt as they tugged at her nipples and bit her in their increasingly vigorous bouts of play. The cubs played when Sonara joined them, usually in the late afternoons or early mornings. When left alone they still crawled into hiding and would remain motionless for long hours while their mother went away to hunt or to be sociable with other pride members around the wells or lake. This response to being without a guardian was important to their survival for mothers could not stay with their cubs all the time, and there were dangers present for unprotected cubs.

One mother seemed especially concerned about protecting her cubs, seldom leaving them to join the others and reluctant even to go hunting. Siku was a nervous mother and kept her two young cubs hidden by themselves in Jasmine Kopje. No other mothers shared the kopje so she had to guard them all alone. Siku's cubs could only just squirm about, safe under the wall of bushes, hemmed in by sweet smelling jasmine vines. But the nook wasn't safe enough for Siku. She frequently left the den to climb to the top of the kopje, scan all around and return to her cubs. If any of the other pride members came near, she would snarl at them. She grew thin and more bad-tempered as the days went by and the cubs caused concern by being able to clamber out of the den.

One night, Sonara, Swala, Shiba and Safi left their cubs and assembled on the flat rock by Fig Tree Kopje before going hunting. Siku descended from her den to join them while the mothers stayed with their younger cubs in the many concealed dens around Sametu. As the hunters were leaving, Siku turned back and stood uncertainly beside Jasmine Kopje, concern for her cubs competing with hunger. Protectiveness won; she climbed back up to her young ones.

The moon was a thin crescent on the eastern horizon when Siku awoke to the sounds of snuffling. She left her den quietly and emerged onto the shelf outside the bush-hidden hole. There in the faint light from the moon was a hyena, advancing through the herbs around the kopje, nose to the ground. Siku lowered herself to the rock and laid her chin between her claws. Her muscles were taught and the tuft of her tail waved an ominous but unseen signal. She sprang. Off the rock and on to the hyena in two smooth bounds, she slammed the spotted night creature to the ground. She shook the hyena, picking him up in her strong jaws and biting hard through the thick neck. The hyena died quickly but Siku held on for a long time. Finally she released the dead beast and lay nearby panting, while the sliver of moon faded and dawn spread across the sky.

Sonara's marsh den

She was still there when Sonara and Shiba strolled by on their way to the marsh at sunrise. They greeted her and sniffed the lifeless form of the hyena. It did not smell good to eat. Then Siku returned to her cubs. Immediately she moved them. She carried each to the marsh, putting them in a clump of rushes between the hidden litters of Sonara and Shiba. The cubs nestled in the hollow of flattened rushes, then crawled between the stems and out of sight. Siku perched vigilantly above them on the matted mound.

Nearby, Sonara lay on the fence of rushes above her own cubs and gently pushed them back into the pen when they tried to climb out. One of Sonara's cubs seemed to have inherited her mother's inquisitiveness for she poked her head into every clump, scrutinized and tasted beetles and tried repeatedly to escape her confines to explore the outside world. Both cubs were very playful and their coordination much better now at four weeks of age. In fact, Sonara found it difficult to keep her cubs in their pen of rushes. When she perched above them they would mew and look up at her reproachfully, often trying desperately to scrabble up the smooth rushes, slipping back into the nest. Trying to get out became a game and the two cubs would leap and struggle, only to fall back, tumbling into one another, wrestling and rolling until they tired and finally crept under the clumps for rest.

The cubs were soon able to escape on their own and Sonara moved them. She carried them one by one to Owl Rocks, depositing them high in the central hollow. There the cubs could clamber about on the dry, soft grass and also explore around the lower rocks as they became more adept. There were many places to investigate and also hiding places in the rocks where the cubs could

73

get out of reach should any danger threaten. Best of all, to Sonara, there was another guardian close by for Safi still had her two cubs in the lower shallow cave of the same kopje. The two mothers kept their cubs apart for only a few days but while they were gone, the two litters managed to find one another one evening and were practically inseparable after that. When Sonara and Safi came home the cubs would scramble out of their hiding place to greet the mothers and then go from one to the other to suck.

Swala also brought her cubs from Fig Tree Kopje and the three litters soon exploited the wonderful opportunities of group life, playing and snuggling and feeding together. If one of the three mothers happened to return by herself, all the cubs would pour out to fight over her nipples, paw and play on her. So the mothers tended to go and come together and shared not only one another's company but the tasks of suckling, playing, guarding and disciplining the active young cubs.

When her cubs were not yet two months old, Sonara found she could not make them stay in the kopje when she left. One morning before sunrise she wanted to go to the wells for a drink and left Swala and Safi lying with the mass of cubs under the owls' tree. But her two cubs would not be left behind and bounded along at her heels. Sonara returned twice to the rocks but each time the cubs simply scampered after her when she turned to leave. If she had hurried away, the cubs might have stayed, but Sonara went at a slow pace, letting the cubs follow. It was time to introduce them to some of the rest of the pride and learn more about Sametu.

The two cubs pranced, jumped and stumbled along behind Sonara. They ran between her legs, leapt at her tail tuft, careened into one another, and generally had a marvellous time. Sonara walked slowly, alert for any danger, pausing when the cubs fell behind. Two hidden pairs of ears tracked her just as carefully, for the steinboks now had the mobile cubs to worry about as well as the other lions. The cubs did not notice the steinboks, they were too busy investigating holes, extricating themselves from the tangled grasses and galloping after Sonara.

Steinboks

At the edge of the marsh, scattered around the wells were most members of the pride. Sonara stopped at the first shallow depression and called her frolicsome cubs to her with a loud "hurmph". They gambolled up, tripping and falling. Sonara waited while they leaned against her legs then lowered her head and repeatedly rubbed each cub with the side of her face, marking each with her own personal scent. The cubs pressed against her massive head, reassured. Three lion lengths away, Shiba sprawled, watching Sonara and the cubs. She sat up as they approached and raised her lip slightly. The hint of a threat sent the cubs tumbling back to Sonara who rubbed them with her head several more times. Then they tottered onwards towards Shiba again. This time, Shiba extended her nose and sniffed the two furry infants. One cub tripped over her paw and rushed back to Sonara's protective presence. The other cub sniffed Shiba's cheek and as she swung her head, the cub rolled onto its back to receive a brief nuzzle before scrambling back to mother.

Sonara greeted Shiba warmly and lay down beside her, licked each cub and then her companion. After a short time, Sonara led the cubs on through the rest of the group, threading her way among the sleeping bodies to the edge of the spring where she drank slowly. The cubs were in the vaguely familiar world of rushes again and started exploring the clumps at the edge of the marsh while Sonara lapped. Over by the spiky trees, Kesho and Kali lazily eyed their two offspring but did not move when the cubs approached. Sonara called her two

adventurers to her urgently, they would be introduced to the big males another day when she and they both felt more secure.

Sonara led the cubs into the marsh to a rush clump near Shiba's litter and spent the day among the rest of the pride. When stars were twinkling over Sametu and the assembled hunters set out, the two cubs followed Sonara at the rear of the group. Sonara had to stop often as the gay cubs tired, and finally she dropped out completely to return with them to the marsh. She stayed with them for a while but when she heard the pride growling over their dinner, she moved off quickly, leaving the cubs sleeping in their clump.

Not long after this initial expedition Sonara's cubs were joined by others on hunting treks. Soon all the litters had been allowed to leave their dens and join other adults and other cubs. Even Siku let her precious two spend some time playing with the other cubs. But she seldom let them follow along on pride manoeuvres. The eight mothers frequently gathered at the edge of the marsh and shared the delights and burdens of cub rearing. All were fairly tolerant of any combination of cubs and usually let any suckle, play and climb all over them. They took turns to guard, fetch or entice the cubs back to hiding places in the marsh or kopjes. Sometimes one mother might be left with ten or more cubs when the others went off to hunt, at other times the cubs might be left entirely alone.

When they had brought all their cubs together, the eight mothers shared eighteeen cubs between them. More had been born, but died at birth or in dens. The eighteen had survived into their second or third months of life, the older ones not much bigger than the youngest. The cubs treated all the adult females as mothers, even Sarabi who had none of her own. But they did learn that Sarabi had no milk to give them though she was almost always a responsive playmate and sometimes a leader and guardian when with one or more of the "true" mothers.

This was a happy time of life for the cubs and for the pride as a whole. Prey was still plentiful and the plains green, decked with flowers and dotted with rain pools. Gradually the cubs became acquainted with their extended world as they followed their elders.

One night seven cubs tagged along on a hunt, bounding this way and that, stopping to wrestle, hide, stalk and chase, practising the motions if not the right sequences for hunts they would perform in the far future. Sonara and Swala kept behind the other females, shielding the cubs from any danger and also protecting the hunters from the interference of the heedless cubs. The energetic brood explored around a group of bushes near the spot where the guardians rested. They got a fright when an undulating black shape separated from the clump and eddied away, the startling mantle of silver over its back making it seem to hover in the bright starlight. It was a ratel or honey badger and a second one followed the first out of the clump, both gliding away silently into the night.

The cubs rushed back to their mothers as the ratels went in the opposite direction. The ratels seemed to have a base in the saddle on the ridge west of Sametu and the older lions had grown used to seeing them shimmering in the starlight. They left them strictly alone. Ratels were ferocious animals, using their sharp teeth to bite and their long claws to dig, rip and tear. They were aggressive if disturbed and smelled musky, a smell that readily deterred lions, especially if they had learned the terrible taste that went with it. Luckily ratels didn't bother with, let alone eat little lion cubs. They went their own way, nosing, listening and digging up their insect food.

Sonara called the cubs as they set out to catch up with the hunters who had gone across the saddle claimed by the ratels as home. Carrying a stick in its mouth, one cub led the others on a merry chase, but the play stopped as the two mothers began to run, leading the cubs quickly along the ridge to where a zebra had been caught. All the other females were thickly clustered around the carcass and the newcomers fought for places. The mob growled and grumbled and the cubs entered into the mood, whining and mewing as they too fought for

Ratels

a small share. At three months they had their full set of baby teeth and could chew scraps of meat as well as lap blood and gnaw on bones.

The cubs got in the way of the adults who were intent on eating as much and as fast as they could. It was amazing that the cubs did not get bitten, clouted or rolled on in the scrum. Every so often, a cub did get caught between the bodies of its elders and would squeak and yelp until released. The adults had the habit of carefully licking what they were about to bite. Thus they had time to realize if it was their prey or one of the cubs before sharp teeth sliced any skin. For their part, the cubs were careful to keep clear of teeth and were resilient to blows from paws. By the end of the meal, though, they were filthy, covered in mud and blood.

Kesho and Kali were usually left with the remains of any carcass and tolerated the cubs sharing the last morsels. But it was their mothers who licked clean the gore-covered cubs when they had finally abandoned the bones to their fathers. After a rest and a grooming session, the pride returned to Sametu, leaving the ridge top to the ratels and the kori bustards.

The kori bustard strutted away but he no longer boomed. His mate had successfully hatched a baby bustard and the mating season was long over. The family was having to search hard to find enough grasshoppers to eat as the plains began to dry up. The ratels were having to go further afield too. The lions were ignorant of these small signs of the changing seasons, for their prey animals were still abundant. They could not know that prey was especially plentiful at the end of the wet season only because it was concentrating before leaving the plains altogether.

The sated lions returned that morning to a Sametu that was crisping to a brown colour, the grasses going to sleep again as the rains diminished. The lake had grown more saline because the water was evaporating, and a flock of flamingoes had arrived to filter out the algae that grew in the salty water. Startled by the approach of the lions, they lifted off the dawn-mirrored lake in pink and black profusion. The lions circled the lake, going to drink the fresh water at the wells and spring. The flamingoes settled at the far end of the lake to resume their head-down sieving of the water but were provoked into flight again by an attacking fish eagle. The handsome eagle, like the flamingoes, was but a temporary visitor to Sametu, often perching in the top of the stunted tree by the wells to eat the flamingoes it caught. There were long red legs and pink feathers under the tree and when the cubs reached it they immediately began to play with these unusual toys, carrying them aloft; rushing after one another to steal the bright prizes.

A few days later the fish eagle left to return to her usual range northwards along a river. Soon after, the flamingoes left too, flying high, honking their way south to a much larger lake in the crater of Ngorongoro where they could feed among thousands of their fellows. The plains offered no big lakes, only small waterholes which were disappearing rapidly. The short grass had stopped growing and the long grass was coarse and dry; over the vast plains there was neither food nor water enough for the thousands of animals. Daily the wildebeest trekked off the plains in long dusty lines and the zebras and gazelles left too. The Sametu lions caught the departing prey; the cubs grew fat on their rich diet of milk and meat while the plains slowly grew lean and dry.

Fish eagle and lesser flamingoes

CHAPTER FIVE

Dry Season
(*Dry season, 2nd Year*)

A little dust devil spun across the brown grass, swirling up bits of leaf and soil. It gained momentum at the bottom of the slope as it skipped across the open places between rush clumps, twisting and rattling the stiff stems. Gaining shape and colour, it brushed over the salt-rimmed shore and whirled gleefully across the dry, caked mud at the end of the shrunken Sametu lake. The miniature cyclone swept up the salty dust on the other side of the lake, hurled it at the tough, wiry grass, and scoured the open rock near Fig Tree Kopje. Rushing over the broad rock, the dust devil then began to stagger; it whooshed in disarray up the slope past Jasmine Kopje, tangled itself amidst the thick dry grass, slowed, paled and ceased its short but lively existence.

Sametu looked deserted. No lions lounged on top of the rocks or by the wells or lake shore. Just a few birds stood sleepily around the muddy remnant of lake, the hot, dry wind ruffling their feathers. The mid-afternoon silence stretched out over Sametu and the surrounding plains, empty and still except for the ceaseless sighing of the wind. The most active things visible were the dust devils but there were less obvious creatures like a warthog that brought her family of four piglets to drink and wallow in the mud opposite the marsh. And the five little lion cubs that scared them away when they emerged from a rush clump in the marsh to lap at the spring water.

Startled by the wathog's alarm-snort, the cubs hid again. The warthog couldn't see them when she stopped and turned to peer with her small eyes set between pointed ears and strange, protruding knobs. Looking down her bristly snout, hooked at the sides of the mouth by curved tusks, she wiggled her flat-ended nose. She could neither smell nor see the cubs as they crouched among

Warthog and crowned crane at the marsh

the clumps, so she slowly returned to the water, her piglets cautiously trailing behind. Two elegant crowned cranes moved aside to let the train of warthogs pass. They eyed another waterhole visitor more warily, a lone silver-backed jackal, trotting rapidly to the far end of the thin stretch of water. The warthogs whirled again and rushed away, tails up, then paused to watch the jackal. They waited for him to leave, then returned once more as the jackal's place was taken by a carefully stepping, watchful gazelle.

Sametu was one of the few places left on the plains that could supply water to the thirstier residents trying to survive the several months of dry season. Warthogs were especially fond of water and liked to cover their rough skin with mud. Their bodies took on the grey colour of the mud and its caked, cracked texture contrasted with the reddish brown strands of their long manes. Gazelles needed water less often but they and topi and kongoni antelopes came every few days, as did jackals and the Sametu lions.

The cubs that were left in the marsh had water to drink. Other cubs were hidden in the rocky fortresses around Sametu and baked in their stony ovens, with no liquids, let alone solid food. Their mothers were often away, trying to cope with what was to them a crisis. Prey animals were no longer abundant because the migrants had left the plains completely. The few residents were hard to catch. The adult females of the Sametu pride had to roam widely in order to hunt. They would have been hard put just to feed themselves, an unusually large group of females trying desperately to survive in a barren land.

82

But the nine females also had to feed their five-month-old cubs, all eighteen of whom demanded more and more meat as their mother's milk diminished. And there were Kesho and Kali as well; the two big males ate an enormous share of any kill.

Survival was at stake, not only for each individual but for the pride as a whole. Must the pride abandon Sametu completely and become wandering nomads, retreating to the richer woodlands where more prey lived? By becoming nomads, the females would put their own lives and their cubs at great risk. The resident prides living in the woodlands and the border of the plains did not welcome intruders, especially during the dry season, and would chase and even kill them. The only other alternative was to try to survive on the plains where there were fewer enemies. However, the lack of enemies was due to the shortage of food; if Sametu had been an ideal area for lions to live, the nine females would not have found it unoccupied.

The Sametu lions didn't actively choose to stay on the plains, they simply struggled to survive. The best hunting grounds were already occupied and defended by other prides; the cubs were not old enough to travel far; so the pride just had to subsist on what it could find to eat locally. As their hunts took them further afield, they often left their cubs in dens at Sametu. The cubs were relatively safe there, though sometimes they went for several days without food. However, they would soon have to travel with their mothers if they were to eat. And travelling with the adults would cause other problems, both for the hunters and the cubs themselves.

Sonara came back to Sametu in the evening when the dust devils had subsided and warthogs and other creatures had left the lake shore. She seemed to have just enough energy to reach the bare, flat rock next to Fig Tree Kopje. There she lay down and called to the cubs hidden up in their fig-covered cave. After some time, three rounded heads appeared from behind an aloe on top of the kopje and peered down. Sonara called again and the cubs poured down the cleft and ran to her, rubbing against her body, jumping at her head, getting licked and rubbed in return. A fourth cub arrived and a squabble began over Sonara's nipples. She lay flat on her side, too tired to resist. The hungry cubs sucked vigorously and their sharp teeth hurt her nipples. Two more cubs descended from Fig Tree Kopje and were crossing the sandy open space when a hyena whooped and they rushed back to safety. Sonara called them to her but there was no milk left to offer, all they got was the reassurance of her company.

The cubs who had got a share of the milk were refreshed as well as reassured and began to play. Their spirits were heightened by the return of Swala and Shiba. These suckled the cubs, who growled and fought over the diminishing milk supply. The mothers never showed any favouritism, so the weaker cubs had to compete equally with their more robust brethren. The stronger and better-fed cubs usually managed to get more, leaving the weak at an ever greater disadvantage. The difference was not yet particularly noticeable, for the cubs were all rotund and bouncy. The real test would come later when the cubs had to travel with the pride.

Dawn came, spreading banners of gold and pink across the sky. The sun was just coming up when Kesho and Kali came back to settle by the wells. Several of the females would not return at all that day for they were far away, hunting as an isolated group. The pride members were seldom together any more, they hunted in small bands, came and went from Sametu only to feed the cubs or for a drink or to spend a rare day of rest on the kopjes or beside the marsh.

Sonara felt the nip of little teeth again on her sore nipples and sat up. She looked over at the lake which reflected the colourful morning sky. Feeling thirsty, she rose and went down the rock, pausing to look up at Swala and Shiba lying on their sides, some cubs still at their nipples. Three cubs watched Sonara turn and leave; this time they would stick with the mobile females. They tumbled down the rock and ran at Sonara's heels, rebounding off her feet, leaping at her legs and tail. Nearing the sliver of silvered water, the cubs rushed ahead. A clatter of wings and repeated chuttering calls of "arr-wrukka, arr" signalled the take-off of a large flight of yellow-throated sandgrouse, sounds symbolic of dry season. A cub ran at a pair of the smaller chestnut-bellied sandgrouse which fluttered up too, and circled to land on a bank as the lions walked around the lakeshore.

There were hundreds of sandgrouse standing on the shore and at the water's edge. They blended beautifully with the muted colours of the dry brown grass, the pale yellow sand flecked with grey and black, all softly dusted by the warm golden light of early morning. As soon as the lions passed, the thirsty birds eddied in wavelets of tan and brown to the water, dipped their beaks and lifted their heads to swallow. More were arriving all the time, the clatter of their wings and continual chorus of quick "garks" filling the air with sound. After drinking and preening, the birds began to leave in curved flights, flapping rapidly away to spread out over the plains in search of grass seeds.

Sonara and her three followers rounded the end of the lake and met Safi and several more cubs playing near the edge of the marsh. The two mothers went to the wells to drink while the cubs bounded along. They were joined later by Swala and Shiba who brought the rest of the cubs from the flat rock. The reunited cubs played together happily while their mothers tried to sleep through the long windy hours, dust devils snaking around them.

The cool smell of evening awoke the lions and they set off early up the slope to Ratel Ridge. Six of the cubs refused to be left behind and followed along

Yellow-throated and chestnut-bellied sandgrouse pairs

behind Sonara, Swala, Safi and Shiba. The clear light of the moon was replacing that of the sun when the hunters reached the saddle to scan for prey. A few kongoni antelopes grazed in a cluster, tan-coloured like the grass and secure in the moonlight, for they could easily see the lions on the ridgeback. Up the ridge towards Bustard Hill a lone topi had his long head in the grass nibbling the few green shoots. His vigilant face appeared every so often to check on the lions. Down the slope to the west, a small mixed band of Grant's gazelles and tommies moved steadily further away, and three leafless poles were really the necks of resting ostriches.

The four hunters waited for moonset, cubs lying between them, playing gently with one another and with their tolerant mothers, or exploring holes in the termite mound on which they all reclined. At last the moon went down and the lions deployed themselves for a hunt. Unfortunately, the topi had joined a group of its fellows and moved off further up the ridge. The beige kongonis were now huddled warily just over the slope in the darkness. Sonara and Shiba crept towards them, stealing from termite mound to termite mound while Swala wriggled low through the tufted grass. Safi remained with the cubs for a while but as the long minutes passed and the group of kongoni began to drift upslope, she too edged out, circling behind her prey.

The cubs stayed where Safi had left them but they were not happy about it. Clustering together they tried to sight their guardians over the clumps of grass. One cub suddenly mewed softly in alarm as a black and silver ratel rippled by in the darkness. The next thing heard was the bonk of hooves striking the ground as the kongonis bounded away. Their keen senses had detected the tiny sounds, traces of scent, slight movements in the grass, and these signs of deadly danger set them off, pronking high, on stiffened legs, warning one another and all around that enemies were near.

The thwarted hunters reassembled at a termite mound. Safi returned to collect the cubs who had ruined the stalk. Greetings and comforting gestures followed, then a short nap before they set out again down the slope to the west into the first reaches of the wide, outer Seronera valley, leaving Sametu behind. The females tried hunting gazelles but to no avail. They continued west and Sonara stopped for a while to dig around a deep hole that smelled deliciously of warthog. While most grazers had to remain fleet-footed and wary night and day, the ungainly but clever warthog slept below ground at night. It used old aardvark or hyena holes or enlarged any other hole that it could find. At dusk, it would back into the burrow, keeping its vulnerable rump well away from the entrance while its sharp tusks faced out, ready to slice any paw or nose that dared enter.

Sonara finally gave up digging alone and loped along to catch up with the others who had paused to watch another group of gazelles. The nervous gazelles kept moving away, not giving the lions a chance to stalk. Towards morning, during that dead time of night when there is no wind and the air is cold, Sonara almost stepped on a crouching hare hidden in the grass and caught it easily. She ate it alone while the others waited for her. Then they went on.

The first strokes of light found the band of hungry lions on the slopes of the valley that led imperceptibly down to the Big Marsh at the outer limits of the Masai pride range. The sun rose over Ratel Ridge behind them as they looked down at a jumble of rocks at the beginning of the valley. The rocks offered little shade, their rounded grey humps only just showing above the long grass. They looked like a string of grazing warthogs with only their backs visible. The lions did not bother to go to the rocks but stretched out in the open, the cubs creeping into the bases of grass clumps.

Drowsy and panting, the lions lay all day in the sun and wind. In the hot afternoon Sonara sat up abruptly to stare towards the warthog-like rocks. Between the clump of boulders and the lions was a little train of moving grey rocks that were indeed real, living warthogs. Sonara tensed, lowering her head and peering over the tufts of grass. The five warthogs had not yet noticed the sprawling lions and continued to crawl along slowly on their front knees, their

86

The kongonis bounded away

faces hidden in the dusty growth. They rootled for fresh shoots, every so often lifting their heads to look around. If all was a safe shimmer of grass, down would go the knobbly heads again and the train of warthogs would munch onwards.

Two cubs awoke and watched Sonara intently. They were still not fully aware of the connection between food on the hoof and food in their mouths but abrupt changes in the adults' behaviour alerted them too. In time they would learn to be still when they saw their providers watching and stalking. Now the two cubs made their way to Sonara, trying to get at her nipples. A third cub joined in the efforts to wedge into Sonara's groin. Sonara swung her head around and bared her teeth, growling low. The cubs stopped their grumbles briefly but as Sonara resumed her contemplation of the warthogs a fourth cub joined in and the squabble was renewed with vigour.

Sonara rose suddenly, cubs dropping from her teats; she swiftly moved off in a low crouching walk, keeping behind grass clumps. Swala was now awake too and saw the warthogs. She crept after Sonara, circling around so that she was behind them while Sonara went obliquely. The two females kept low in the grass, flattening themselves whenever the warthogs stood or looked up. Shiba was watching, keeping track both of her companions and the warthogs. For many minutes she watched; then crouching low, she too slunk through the grass. Swala had crept close to the oblivious grazers and was holding very still. The warthogs were methodically munching their way along into the wind and neither saw nor heard the lions approaching.

Suddenly two raised their heads at the same time. Swala and Sonara flattened themselves into the grass. The warthogs had heard something unusual and gazed all around, staring hard at the rocks down-slope. At last they resumed grazing. With great patience, Swala and Sonara let them settle down, then once more started to worm their way through the grass. Shiba slithered along on her belly into position between her two sisters. Once again the hogs lifted their heads, all of them this time. They snorted in alarm and wheeled away, tails up, stiff signals of distress. They trotted off a short distance, then turned to stare – not at the hunters, who were hidden flat between grass clumps, but at Safi, who was sitting bolt upright because she had been awakened by the cubs biting her nipples. The warthogs peered at her, then swung their knobbly heads to stare at the rocks. A tawny shape could just be seen at the base of a boulder and the distant sound of a sneeze came to their ears.

Off they ran at a gallop, tails straight up, tufts waving. They stopped further away and looked at the two groups of lions, now sitting up watching the retreating meal. Up went their tails again as they trotted away, their taut fat rumps bouncing above their little legs.

Sonara and Swala looked at each other. Occasionally they managed to catch one of the succulent beasts, so it was always worth a try. The presence of the cubs obviously made hunting much more difficult. The females did not show any sign that they resented this, returning to the cubs, nuzzling and

Warthog

licking them after they greeted one another. Then they all looked over at the rocks as another sneeze came to them. Sonara went closer and saw that it was Salama; also in the rocks were Sukari, Sarabi and four more cubs. In the enervating sun, the females and cubs all greeted one another, then sought refuge from the heat among the nearby rocks. The afternoon drifted away on the drying wind, the sky filling up with busy commuter clouds hurrying westward over the plains.

Topi

When the seven female lions left Warthog Rocks in the evening they moved off quickly and quietly, leaving the cubs behind. Tonight they had to catch something to eat. Along the flank of Bustard Hill they spied a lone topi. They rested and waited until the moon went down, then began to stalk the sleepy old male. He moved slowly up the slope and stood near a big termite mound. Sarabi and Safi carefully crept through the grass behind him while Sonara and Swala went more directly. The others took up various positions along his probable line of escape.

While the topi had his long face down, the hunters advanced quickly, slithering through the grass. The ripples ceased whenever the topi lifted his head. He stayed alert for long periods and slowly ambled along. The great constellations arched overhead as the lions slowly converged on their victim. The topi caught a sudden whiff of lion scent and snorted. He whirled to run, and met Sonara. She leapt at his throat. The old male stumbled when she hit him; and both rolled over as they fell on to the ground. Sonara tried to renew her grip on the topi's throat while he staggered to his feet. He reared but fell again as Shiba caught his rump. Sonara held on to the throat and the topi died quickly.

Swala and Safi ran up and began the quick work of opening the carcass. The others joined in and soon there were grumbles and growls spreading wide as

the hungry females fought for places and portions. The sounds summoned Kesho and Kali who had been sitting on top of Ratel Ridge waiting for just such signals. They trotted along quickly, filled with appetite, intent on getting a good share of the kill.

Sonara managed to tear a chunk off the topi carcass and ate her piece quickly away from the mob before leaving to fetch the cubs. She saw Kesho and Kali break into a run as they neared the group. Females fled in all directions, taking whatever they could carry, leaving the rest as tribute to their hungry males. Sonara went down the slope back to Warthog Rocks. The night was ebbing and the air cold as she neared the cluster of boulders, calling the cubs. One by one they came, tumbling out to greet her. She nuzzled and licked the first few, then turned to lead them back to whatever might remain of the meal. She was still hungry and hoped there might be a little morsel for her too. But it was a long way and took a long time, with the cubs straggling along in a slow reluctant line.

When the flock and shepherd arrived almost nothing was left. Kesho and Sarabi lay side by side, clinging to the head and neck of the topi skeleton, grumbling low at each other. Neither would relinquish claim to the remains and each waited to exploit a moment when the other shifted and lost a grip. The deadlock was broken when Kesho tugged especially hard and wrenched the bulk of the carcass away from Sarabi who clung to the head until it separated. She rushed off with it just as the cubs rushed in, swarming over the remains

that Kesho tried to keep to himself. But the hungry cubs ignored his growls and soon were squabbling among themselves over the scraps.

Sonara got a sharp slap and ferocious snarl from Kesho as she too tried to wedge in with the cubs. She nosed around but gleaned nothing so went to join the others. They were all still hungry, lying here and there cleaning themselves as the clear sky filled up with another dry, daunting day. None of the lions noticed that not all the cubs they had left at Warthog Rocks were present. Lions cannot count much above two and it was virtually impossible to keep track of the large numbers of cubs as the mothers mixed and moved. If any cub was too weak or unobservant to respond to a summons or straggled too far behind on a trek, it could get lost and abandoned and would surely die. The cubs left at Sametu were little better off for they too were exposed to drought and starvation. And all the cubs, both those hidden in dens and those tagging along with the pride increased the burdens on their providers by necessitating long expeditions to fetch and feed, as well as exposing themselves to dangers when their guardians were absent. The burdens and dangers increased as dry season wore on; thus the numbers of first born cubs of the Sametu pride began to diminish.

The topi was the last large shared meal the lions had for several days. Their hunts led them further and further down Warthog Valley into the Seronera watershed. The many creases and folds of that wide, dry basin converged upon one dark green stretch of rushes, the Big Marsh on the outskirts of the Masai pride range. The Sametu females were somewhat familiar with the area, having accompanied their own mothers there during past wet seasons. Now, even though it was dry, there seemed to be more prey in the Seronera basin than around Sametu. So the females hunted around the four low rounded hills that bordered a long shallow valley to the west and led the cubs on tiring treks across the gentle slopes and ridges. Meals were few and far between and all were growing thin. It was a relief for mothers and cubs alike to have the water and cover provided by the Big Marsh. The marsh was a haven only because the Masai pride never came there during dry season and the Loliondo males seldom patrolled that part of their range. Nevertheless, it was a dangerous spot and the Sametu females went carefully and silently as they stalked the scattered prey round about.

One night, Sonara, Swala, Safi and Shiba left their cubs in the marsh and went hunting along the side of the first hill. They stopped to scan for prey near an oddly shaped termite mound, huge and very old, a squarish shape with a big hole at its base. It looked like a friendly monster, a troll or gnome guarding the flank of the hill. From there the group hunted gazelles and kongonis in the saddle between the first and second hills. The hunters became separated during the night and Sonara found herself alone. She returned to the marsh at dawn, passing the troll-shaped mound, where a mysterious dark head pulled back into the hole. Always one to stop and investigate, Sonara sniffed around but discovered nothing. She went on back to the marsh, where she lay down on a low bank to call for the cubs.

Five muddy, wary heads had appeared from between the clumps of rushes when a call sounded almost at her shoulder. Sonara rose and turned immediately, ready to defend herself and the cubs, but it was only Siku, leaping on her with such joy that Sonara collapsed sideways, almost crushing a cub. Siku had been tracking her missing companions for two nights, hungry, thirsty but above all, lonely. She was torn between returning to her weak little cubs back at Sametu or searching further for her hunting companions. The stronger scents at the marsh had lured her on and then she had heard the sound of Sonara's calls.

In her dash across the mud flat and around the marsh, Siku had not noticed the huge paw prints nor smelled the warning odour that surrounded them. Eagerly she had run to join Sonara who just as eagerly greeted her, neither aware of an approaching figure. Behind their screen of rushes the two females licked each other happily, cubs emerging from their hiding places. A loud low grumble shocked all ears; the cubs vanished back into the rushes and Sonara and Siku bolted for cover too.

Lengai stood on the far side of the mud flat, water oozing slowly from beneath his massive paws. The wind fingered his long mane and rattled the stiff stems of the rushes like dull swords clashing. He stared around the shimmering expanse of rushes and slowly marched along the edge of the water. Each paw was placed firmly as he paraded around the marsh. Arching his great neck,

head high, sniffing, he strutted through the long grass and across a gully to stand on a bank above the marsh. The rushes revealed nothing to his cocked ears and eyes squinting against the bright sun so Lengai began to sniff all over the area. Slowly and carefully he quartered the grass, sniffing and frequently stopping to scrape.

The sun was high when he finally finished marking and inhaling information. He stood watching the marsh for a long time, then turned back to the gully where a little bush offered some sparse shade. Lengai reclined facing the marsh, lowered his shaggy head, rolled on to his side and dozed.

In stealth, Siku and Sonara emerged from the far side of the marsh in the corner furthest from Lengai. Siku was agitated because the terrifying appearance of the male had reminded her that her own cubs were alone and unprotected at Sametu. Sonara followed Siku rather reluctantly. Having a close companion in the face of danger was preferable to staying at the marsh alone and she could not call out or search for the hiding cubs until Lengai left the area. So Sonara accompanied Siku who led her at a fast pace, heading east along Warthog Valley. Halfway along the route two white-bellied bustards rushed out of a bush clump, causing the two lions to pause and catch their breaths. Sonara almost turned back then but Siku's determined manner and the promise of a drink lured her on.

Siku and Sonara walked steadily, despite the heat, gradually ascending the sun-scorched slope that led them away from the marsh to Ratel Ridge. They

White-bellied bustards

stopped to pant in the heat increasingly often, and near the pile of rocks at the head of the valley they detoured to hunt warthogs. But Siku was impatient and spoiled the hunt, rushing too soon. On she went; Sonara followed. They paced eastwards through the shimmering heat and dry, dusty grass. Crossing Ratel Ridge, Sonara paused to gaze down to the shining patch of water at Sametu. She was very thirsty and so was Siku who drew ahead, her pace accelerating as she neared home, water and her hidden, helpless cubs.

Whirls of alkaline dust saluted them as they crossed the exposed outer crust of the ever-diminishing lake. Blinking, at the dust and glare, Sonara and Siku bent to lap the salty water, too tired and thirsty to go further. Sonara collapsed on to her side at the water's edge but Siku rested only briefly and then continued, going directly to her cubs hidden in Owl Rocks. The steinboks watched her disappear among the rocks. Later she passed by again, having given what little milk she had to her two emaciated cubs.

Siku returned to Sonara, still lying exhausted at the edge of the lake. She rubbed her face gently against hers, then collapsed alongside. They slept side by side for what was left of the hot afternoon. When the wind dropped and evening stole quietly over the brown plains Sonara rose and tried repeatedly to get Siku to join her on the long trek back to the Big Marsh and the cubs. But Siku would not leave and finally Sonara moved on, stopping often to call back to her unresponsive companion. She investigated a hole that a hyena had slipped into, then went on, up to the top of Ratel Ridge and over the other side.

Sonara rested briefly on the ridge crest in the starlight then dropped down into Warthog Valley. As she neared the rocks she saw two silent shapes and stopped dead, staring intently at the mounds that must be males. She realized it was Kesho and Kali and carefully peed and scraped the soil where she stood, leaving her mark in case they should wish to know who it was who passed them in the night. On down the valley she went, retracing her steps of the day. In the early morning she arrived back at the marsh, going straight into the rushes, quietly looking, sniffing, listening for any sign of the cubs. She came out the other side not having found a single one, still calling low and moving carefully, senses fully alert for danger. But no lions answered, big or small, friend or foe. Lengai had gone; so had the cubs.

Wearily, Sonara searched further, sniffing her way up the flank of the first hill. She came to the gnome-shaped mound and caught a trace of scent. Calling repeatedly, she finally saw four sluggish cubs emerge from a grass clump to come stiffly forwards. Sonara sighed and lay down, licking each cub as it rubbed under her chin and against her tired body. Again, not all were there. Sonara and the other mothers who had kept their cubs in the marsh would never know what had happened. The missing ones may have been killed outright by one or more of the Loliondo males or simply lost, to perish from thirst, hunger or other causes.

Sonara stayed with the cubs all day and by evening had recovered from her exertions enough to greet Swala, Safi and Shiba when they joined her and their cubs at the gnome mound. None of the four mothers had much milk to give to their four surviving cubs and within a few weeks, the cubs were completely weaned, totally dependent on what they could get to eat at the infrequent kills. By the time the rains had returned to the plains, only three cubs remained out of the eight litters born to the Sametu mothers. Four mothers lost their cubs entirely, including protective Siku. She, Sukari and Salama mated again at the end of the dry season, starting anew. Lions have no breeding or birth season; they let the environment pass judgement on what time is best for cubs to survive.

Luckily, all the adults of the Sametu pride survived that first full dry season on the plains. The hunters, producers, protectors – the mothers and fathers – were all alive by the beginning of the rains and had even managed to keep three cubs alive as well. The Sametu pride continued to hold and defend a range; they qualified as a resident pride, coping with the dangers of dry season and ready to enjoy the abundance of the long rains.

CHAPTER SIX

Wet season
(Wet season, 3rd Year)

The storms crawled again over the wide grasslands, leaving glistening trails that turned to green within days of their passing. With the spicy lure of wet dust and rain in their nostrils, the migratory animals poured out of the Serengeti woodlands to flood over the plains. Waves of wildebeest, gazelles and zebras eddied about the rocks and gentle valleys where the lions roamed. Surrounded by this tide of life, the Sametu pride regained strength and solidarity.

One morning in early wet season the lions greeted the herds and the dawn with resounding roars from the heart of their range. Nine adult females, two big males. The three cubs added their squeaks to their parents' more glorious thudding roars. The one who tried hardest, though her attempts came out as plaintive miaows, was Sisi. Sisi reclined like her mother, Sonara, on her belly, head up, already a miniature lion at ten months of age. She was slim, alert and neat, her smooth coat a pale sandy gold. Near Sisi sat Susu, Swala's daughter. Though the two were the same age, Susu was bigger, a rotund, bulky cub. She sat upright, her belly curving over her fuzzy feet, adding only a few small squeaks to the whirlpool of roars. Susu was a dusky fawn colour, darker than Sisi; she was fat, happy and lazy.

Between Sisi and Susu lay Sam son of Safi, hints of his maleness cresting on the back of his thicker neck, and with a fluffy bib of white hair on his chest. Sam gave a few squelps as he tried out his pathetic roaring voice, then rolled to one side to sink his terrible milk teeth into Susu's throat. Susu retaliated with kicks at Sam's belly as they wrestled while the last trickle of grunts echoed softly from the Sametu rocks. The symphony was hushed as the grunts melded into the murmuration created by the host of other creatures living on the plains:

Sisi, Sam and Susu

warblers zit, zit, zitted invisibly in the air, plovers clinked, larks sang, insects buzzed, gnus honked, zebras barked, bustards boomed, jackals yapped, hyenas whooped. Upon this ebb and flow of sound came distant roars from the north telling the Sametu pride that three Loliondo males were with the Boma pride, edging to the southern limits of their range as they too were lured further along on to the plains after the vast herds.

Everywhere animals were on the move. Some had even travelled much further than the hoofed herds to reap the bounty of the plains. White storks soaring overhead had left their nests on the rooftops of Europe, and swallows which had lately skimmed the summer lawns of England now hawked for insects over warmer, wilder pastures. In the wake of the eager grazers came other hunters, from the egrets and yellow wagtails that trailed the herds, to the hyenas, jackals, cheetahs, wild dogs, and nomadic lions.

Nomads trespassed on the Sametu pride's reserves and adjacent prides also encroached on the northern and western boundaries. The Sametu lions kept their territory intact by roaring often and leaving scent marks up and down the valley and across the ridges. Kesho and Kali patrolled frequently, chasing off any intruders. The majority of the pride could often be found in the centre of their range where Siku, Sukari and Salama had hidden their new litters. With their range protected and prey plentiful, the Sametu lions had a brief season of luxury and leisure before the rains ceased.

Cape buffaloes

100

This period of relative ease provided free time to indulge in prolonged bouts of play. The adult females were still young, an exceptionally merry group of five-year-olds who would gambol and romp with each other as well as with the cubs. For Sam, Su and Si, the play sessions were an important part of their upbringing, allowing them to learn about both social and hunting skills as they stalked, chased and fought one another and the adults. They playfully hunted other animals too and slowly learned what was acceptable prey and how to catch it. The cubs would need two years to master their lessons and the adults were still learning.

On that day the lions rested near the wells. They awoke in the evening to see a black block of beasts advancing down the valley. Thirteen buffaloes, old and young bulls, lumbered slowly along in the twilight. One by one the lions sat up to watch the bulls coming their way. The adults had seen buffaloes in their youth and some had tasted their tough but delicious meat, but to Sam, Su and Si the big beasts were a novel sight. The buffaloes clumped along the shore of the enlarged lake, becoming dark silhouettes as the sky passed from pastel pink to dusky grey. They coalesced into a solid black mass between the end of the lake

and the sentinel acacia where they munched the grass with heavy indifference. The lions ranged themselves along the far bank while a maiden moon beamed down from above, making the buffaloes seem even blacker in contrast to their moon-bleached surroundings.

None of the lions seemed to be stalking in earnest. Sonara led some among the grass clumps towards the buffaloes but Swala and Sarabi sat in full view on top of the bank; the remainder of the pride watched from around the wells. Sisi could restrain herself no longer and crept away from the audience at the wells; she crossed the open grass in a crouch and disappeared over the bank to join Sonara. Sisi showed clear signs that she would learn to be a good hunter for she was always keen, careful and patient. She had already caught small animals such as hares by herself and even participated in communal hunts without interfering. Sam and Susu were usually tardy if they showed any real interest in group hunting at all, but they vigorously joined in once the prey had been despatched.

Sam and Su watched Sisi until she vanished, then openly walked over to Swala and Sarabi on the bank. Their movement alerted one buffalo who swung his massive head up with a snort. None of the lions except Sonara and Si made any effort to hide. The rest of the buffaloes also raised their heads and stood sniffing carefully, their wet, leathery noses shiny in the moonlight, their lashed eyes peering at the grass clumps. With a gleeful bound Sonara and the others rushed at them. The buffaloes snorted and whirled away en masse to stand tightly together on the lake shore, turning to stare at the pale shapes emerging and disappearing in the grass.

More females joined Sonara and the others as they crossed through the rushes and made another charge at the buffaloes. The great beasts advanced towards the lions but suddenly turned and thundered off, not stopping again until they were well away. Then they turned back to stare and toss their heads from side to side to get a better view. The buffaloes didn't seem particularly afraid; the lions, for their part, had little wish to pursue the alert, huge-horned and powerful prey. They stood watching on the moonlit shore until the entire pride had assembled. One more charge and the "game" was settled when the buffaloes decided to retreat down the valley, ponderously and in no haste.

Tattered clouds cloaked and uncloaked the moon as the lions played along the lake shore, dashing, leaping, and bounding about in the highest spirits. Kesho and Kali sat apart, licking themselves and waiting for the real hunt to begin. Finally the hungriest females headed up the slope to Ratel Ridge, followed by the rest, frisking along. Playing their way up the slope they reached some termite mounds and settled down to scan for dinner.

In the feeble moonlight from behind thickening clouds, the lions saw a few topi, gazelles, some patches of zebra and a far off herd of eland. Best of all was a nearby band of young gnu males, most of them yearlings recently rejected by their mothers who were soon to have calves further out on the plains. While the hunters scanned for food, Sam, Su and Si watched the two local ratels trotting along, their black and silver coats beautiful in the gentle moonlight. Suddenly they seemed to vanish as the moon slipped into a solid curtain of clouds and departed for the night.

It was time to hunt seriously. Several of the females were already off the termite mounds, quickly moving away through the grass towards the gnus. Sam, Su and Si stayed put, listening carefully, trying to catch sight of the pairs of black ears that showed where the stalkers were hiding. Before long there was the sound of running hooves, strangled bleating and the onset of growls. Su led the way as they ran towards the kill. When they arrived they found that the hunters had managed to catch two gnus who had stupidly run right at the lions as their brethren fled the other way.

Two tight clusters of lions circled the carcasses, centres for rumbling eruptions of growls. Before the first patters of raindrops came the dung beetles.

Dung beetle

It was almost magical how these black beetles appeared so soon after a carcass was opened and the stomach contents spilled upon the ground. The vivid stench of fermenting grass which came from the ruminant's stomach probably called them as irresistibly as the sound of the kill called the lions. A huge dung beetle landed close to Sam's paw and started gathering her ball of half-digested grass. In it she would lay an egg that would hatch, consume the ball around it,
reform into another beetle and live to roll another ball in its turn. She buried her ball with its secret egg in the damp earth as the rain began to fall more heavily. The beetles worked hard, clearing off the piles of dung while the lions ate.

Next to arrive were the jackals, who stayed respectfully out of reach while the lions scrummed around the carcasses. When pieces were being pulled off and carried away, the jackals began to dart in closer, snapping up bits of meat, scraps, little bones. Both silver-backed and golden jackals were there, though the silver-backs seemed the bolder around the lions. Jackals scavenged what they could from kills made by other predators but mainly relied upon their own skills as hunters, catching mice and small animals and even eating fruit.

Beyond the circles of lions edged with jackals, the hyenas lurked patiently on the perimeter. They didn't risk the slaps or bites of the bigger lions for they weren't as quick on their feet as jackals and were easier as well as bigger targets. Like the jackals and lions, hyenas hunted most of their own food yet were never unwilling to scavenge if they could. Scattered over the plains, skulking around in ones, twos or more, hyenas were always alert to the whereabouts of food.

Near Sametu, a clan of hyenas had a den where each female kept her two black cubs down a hole. Unlike lions, hyena cubs stayed in dens for a long time and only came to kills after many months of feeding on their mother's rich milk.

The hyenas were waiting not so much for meat as for the bones that the lions left. They had the teeth and muscles to crush up large bones, plus a digestive system that could process this seemingly inedible stuff and at the end produce pure white mineral droppings which were deposited as markers around the hyenas' ranges. They milled about in the background, twittering and growling at one another, an agile and clever lot – undoubtedly the most adaptable and economical of the carnivores. Hyenas, jackals and lions were almost hidden by the driving rain that spread a blanket across an already dark night.

Dawn came behind the storm as lions dispersed with their pieces; hyenas skulked about, snuffling, and jackals jinked away with scraps. Sonara half-heartedly scraped a little earth and grass over the entrails from one carcass then abandoned it, having eaten her fill. The dung beetles flew off; they and the rain had mostly disposed of the unwanted parts, blood and dung now safely buried in the soil to feed generations of grubs and grass. The sun rose into a pure blue sky to shine on the last of the diners. To the early morning observer it might look as though one lion in particular was hogging the paltry remains. Kali had been the last to arrive during the night and was still hungry, gnawing on the neck bones, growling at the cubs who chewed at the ribs. Kali hung on to the meatless skull with tooth and claw, rumbling low repeated threats, ignored by the cubs. Two nimble hooded vultures descended as Kali tugged at the skeleton. Hooded vultures were always first to arrive in the morning because they were small enough to fly without the updrafts that came with the sun's warmth to lift the bigger vultures. These two pecked around close to the lions until Sam chased them off and returned to find that Kali had dragged the remains away, leaving only a few bones behind.

Later a few white-backed vultures, one lappet-faced and a Rüppell's came to join the hooded vultures but soon left to search for meatier morsels. Nothing remained of the two carcasses by mid-morning except the skulls which had

Hyena, silverbacked jackal and golden jackal scavenging

been abandoned by lions and hyenas alike. Eventually the last in the chain of eaters would arrive, the hornworm moths. On the horns they laid their eggs which would later hatch into caterpillars and devour the sheaths. While eating their tough dry meal, the larvae would protect their soft bodies with twig-like tubes of horn fragments. Finally only the bone core would be left to be whittled away by sun and rain into food for the grasses.

The lions left long before the hornworm moths arrived taking one last remnant of the gnus who were now part of so many other living creatures. Si and Su had found one of the tails lying coiled like a hairy, fringed scarf and each tugged at an end until Sam joined in to make it a three-way war. Sam won, triumphantly dashing down the slope after the departing lions, his playmates right behind.

The three cubs played and grew as the rainy season continued, surging with life. The rains brought life to the grass which in turn seemed to urge many animals to produce yet more life. The most spectacular producers were the gnus. Births were timed to occur during the season of abundance while the herds were not migrating long distances and mothers and new calves could benefit from the rich grass abundant only during the rains and only on the plains. The gnus produced their calves in a sudden torrent in the middle of the wet season; hundreds of thousands of them giving birth within the short space of a few weeks. This took place south of Sametu, so the pride could not take advantage of the new-borns. They continued to hunt the old, the weak, the

solitary, the inexperienced or the stupid, thus contributing to the evolution of quicker, stronger or cleverer creatures. Culling the new-borns was a job for the few hundred predators and scavengers on the southern plains. One of these was Nafasi.

Nafasi lay on top of a flat rock above the short green grass stretching to the horizons, half-open eyes lazily watching a group of gnus grazing. A few creamy-brown calves gambolled around their chocolate brown mothers whose long beards caught the early morning sun like white banners. He raised his head as the group of gnus honked and wheezed, briskly trotting sideways as they assembled to stare and snort at a pack of wild dogs.

Five mottled yellow, black and white dogs wandered among the gnus, seemingly aimless. They were anything but that, for it was their morning hunting time and they had fourteen voracious pups plus a hungry mother to feed back at the den. Wild dogs are careful hunters; these tested the herd to find the easiest individual to catch. They picked out a new calf, just able to coordinate its long spindly legs. The wildebeest mother kept trying to put herself between the dogs and her wobbly calf. He had been born only a few hours earlier and was already able to run, however, he still needed time to get to know his own mother so that he could follow at her flank and not get lost among the myriad confusing replicas of herself. She edged backwards into the milling mob, hoping to conceal herself and baby in the throng. The wild dogs began to trot towards the group. The gnus turned, beginning to run. The dogs kept on, testing to see if the calf would lag behind.

Nafasi sat up and watched intently as the gnus broke into a gallop, stampeding away, tan calves close to the sides of their mothers. Two of the dogs accelerated to a fast run and circled to the side of the fleeing herd. The young calf they had selected was running exceptionally well and kept tight to his mother's flank. Another calf, however, was trailing behind its mother at the periphery of the group. The lead dogs chose it and cut across the herd to single it out, running hard. They were passed by the other three dogs who closed in on the calf, put on a burst of speed and nipped its heels. It stumbled and bleated. Its mother whirled round to run back towards her struggling calf, tossing her horns at the dogs who ignored her as they killed the calf. Two dogs then nipped at the mother too and she fled, unable to help her dead offspring. It had been weak at birth; she would run now and try for a healthier baby next season.

Nafasi was off his rock and loping in heavy lion style towards the dogs. Before he was even half the distance they had finished bolting down chunks of meat from the small carcass and two hyenas were already on the scene. As Nafasi neared, the dogs broke away, and made off in the direction of their den, the packets of meat in their tight bellies to be delivered to the waiting mouths. Neither lions nor hyenas could carry food home in their bellies and regurgitate it at will as the dogs could. Nafasi arrived panting at the site of the kill just as the last hyena and jackal departed with the remaining scraps. Nafasi nosed around, then ambled back to his rock. It wasn't the first time he had missed getting the lion's share.

106

Birth of a gnu calf

Nafasi wasn't particularly hungry; life on the plains was good even if he had to compete with other predators for meals. He was in his prime, a remarkably unblemished four and a half year-old, his long mane luxuriant, his golden coat sleek and healthy. He was glad to be living again on the plains. Compared to the wooded areas of Serengeti where he had had to spend the dry season, the open grasslands were heaven for a nomad lion. Here there were no thorns to wound his paws, no biting tsetse flies, no hazards like mean old buffaloes lurking in thickets, rhinos in ravines, or men seeking to take his skin and claws by spear, gun or snare.

Nafasi ascended his rock, lizards scuttling off into cracks, and stood on top. He looked for his friends but saw no other lions in the tranquil scene. Nafasi had met a pair of nomads a few weeks after returning to the plains. One was a male, Nasibu, and the other a female, Nani. Nafasi had come to know Nasibu and Nani well, mainly because Nani seemed regularly to want to mate and always chose Nafasi over her other companion. The trio of nomads had been wandering around together, avoiding the territorial prides like the Sametu, and keeping to unoccupied areas further south. Last night, however, they had been chased from a stolen kill by a small pride on the plains that considered the nomads intruders. Nasibu had an uncanny knack for keeping close to Nani and had managed to follow her as she fled, but Nafasi had gone his own way. He would find them, or they would find him no doubt.

Nafasi reclined on the rock with his face into the breeze, chin on paws. The quick movements of the jackals darting among gnus attracted his attention. A couple of females were in labour, producing the morning's latest calves. Gnu mothers had their babies before noon; the young need the daytime to dry, to get their awkward gangly legs under control and to learn the idiosyncracies of their mothers. The jackals were hunting for the afterbirths that came some while behind the calves. These rich delicacies were snapped up quickly. The mothers of new calves kept a close watch on the jackals to ensure they didn't try to take a calf as well as a placenta. Occasionally a mother gnu would lower her head and charge after a jackal, but she never went far from her calf.

The young calves bucked, lurched and frolicked around, gaining essential experience and control. Nafasi could not even have kept up with the youngest of the healthy calves. If he had tried to catch a female unawares whilst in labour

108

she would simply have run off with the calf half-born, somehow pulling it back inside far enough to outrun the lion and have her baby in safety later. Only the fast or long-running predators, like hyenas and wild dogs, really capitalized on the brief flood of gnu babies; or the jackals, indirectly, by taking up the afterbirths and stillborns.

Nafasi closed his eyes, lulled by the warmth and murmurations from the gnus and other animals. He slept soundly as a big blue and pink lizard came closer and closer, hunting flies. The agama snapped up a little fly on Nafasi's big paw, then bobbed up and down, his vibrant pink head glowing in the hot sun. A quick run across the rock and he caught another fly near Nafasi's nose. The little flies he was catching were especially abundant during the wet season, accompanying the vast herds of wildebeest and zebra, clustering on their flanks and around the moisture near the eyes and noses. There were many strays settled on Nafasi and the lizard rushed about picking them off, even daring to scramble on to Nafasi's back. A tempting movement in the furry carpet caught the lizard's keen eye and his tongue automatically flipped out to catch what for

Wild dogs at den

Lion-fly

him was something unusual. It was the tortoise-shell coloured fly peculiar to predators, a special fly that crept crablike into their fur, living off a little blood but not seriously harming its host. Surprised to taste the tough prickly-legged fly, the lizard spat it out.

Nafasi moved and the startled lizard leapt into the air and scampered away into a rock crevice. When the lion rolled on to his side and slept on, the agama came out of his crack and cocked his head. His colours had faded in fright but soon brightened again in the warm afternoon sun. Now he stalked lizards instead of flies. Small dark babies moved out of his way and the grey females curled their long tails over their backs as he approached. The agama rushed about his kingdom checking out all the lizards, had a fight with a neighbouring territory holder, ate some more flies and was nestled safely in his crack, daytime colours fading to grey, when Nafasi awoke in the dusk.

He had heard something. He sat up, yawned and looked out over the plains. The sky was darkening early with massive storm clouds. On the gloomy horizon, a large group of gnus was galloping over the crest of a low rise. Their honking and the thudding of numerous hooves on the short grass were what had roused Nafasi. He peered intently at the other dark forms hunching along after the running wildebeest. The group of hyenas ran with endurance and power, if not with grace, and finally caught a lagging gnu, made heavy and slow by her unborn calf. Soon the dusk was splotched with the sounds of growling, chuckling, giggling as the hyenas began to fight for shares.

Nafasi stretched, arched and moved off the rock in haste. This time he hoped to have something substantial to eat. He began to run, stimulated by the increasing frenzy of the mob of hyenas who were squabbling enthusiastically over the carcass. As he neared the mêlée, Nafasi slowed to a walk. Dark figures roiled amid shrieks, growls, screams and wild laughter. Only a large number of hungry and excited hyenas could produce such a cacophony. Nafasi looked around, perhaps vaguely hoping that his two friends would appear. With them he would not hesitate to barge right into the devilish mass. Alone he was more

cautious. He couldn't risk
getting wounds that could in-
capacitate him seriously, for
hunting needs an intact lion.
Suddenly the hyenas began
to disperse, scuttling off with
bones and pieces of meat.

Hyenas ran past Nafasi,
sniggering and grumbling, as
two lions loomed out of the
darkness going straight to the
remains of the gnu. Boldly, Nafasi hurried to secure his share.
Nasibu and Nani greeted Nafasi
with snarls but permitted him
to wedge in and chew off what
he could. Not much was left.
Hyenas were notoriously efficient at clearing up after
themselves and could be heard
chuckling in the darkness as they crunched their bones. Slicing and tearing
with their side teeth, the much slower lions carefully peeled pieces of meat from
the bits that the hyenas had been forced to abandon. Several bites and swallows
later Nafasi ended up with his teeth sunk into the neck while Nasibu hung on to
the head. While the two males growled tensely over these relatively inedible
prizes, Nani cleared up any scraps.

Nasibu was slightly smaller than Nafasi, his mane fleecy and short, but he
was strong. Nani was a thin, sandy-coloured lion whose knobbly vertebrae
protruded all along her skinny back. This was her second year as a nomadic
female, not an easy life, but one she was stuck with. She had left her home pride
with her litter-mates, all males, when they had been forced out at the age of
two. She no longer had the option to rejoin her former pride for they would not
now have her back: she had been away too long. Nani had roamed during the
first wet season with her gang of brothers but gradually the group had broken
up and she was left with Nasibu. He was a friendly parasite living off her kills,
but a big help when the pair scavenged. Unfortunately, Nasibu was too familiar
a companion to be attractive as a mating partner, and Nani periodically became
interested in mating. She was a barren female, for some sad reason doomed
never to bear cubs but repeatedly trying to become pregnant.

Nafasi and Nasibu broke their deadlock, each male parting with a portion
and walking off to chew on the bones alone. Nani waited patiently for the males
to finish. She had never tried hard to get away from them even though they
inevitably ate large shares, leaving her thin and hungry. Their company,
protection and help were desirable, their cost acceptable.

Rising, Nani pussy-footed over to Nafasi and lingered near him. She had to

112

The nomad trio

compete with a bare neckbone for his attention; he persisted in gnawing on it while she delicately edged around him, flicking her tail. Finally he raised his shaggy head and sniffed deeply, grinning at the scent of the waiting female. Nani paused to pee and scrape with her back paws, slowly and deliberately, then ambled off in her lanky, lean way. Nafasi abandoned the old bone for the pleasures of the flesh. The two wandered off and settled by a waterhole to mate the rest of the night and the following several days and nights.

Nasibu was forsaken when Nani and Nafasi went their way. He depended on his "sister" for company as well as food but knew he was not welcome when she was in the mating mood. He hung around the courting couple, looking like a moping outcast, as indeed he was. At night he wandered around, intent on noises signalling food, but he was an inept hunter and not good at scavenging alone either. During the day he would return to the waterhole where Nani and Nafasi lay side by side as if permanent fixtures. Near the muddy, salty pool was a cluster of spiky, stunted trees under which Nasibu might crawl for a smidgen of shade. A community of spiders had laced their intermingling webs all through the thorny branches, creating a delicate umbrella which shimmered in the heat like spun glass. Slow-witted and uncomfortable, Nasibu would lie like a withering clump of grass in his bower waiting for the evening, hoping Nani would be ready to hunt again. The spiders scrambled about their gossamer homes whenever Nasibu moved, then bounced their way back into position when he was still. There they would wait for the flying prey that tangled itself in their town of traps.

Leaving his silk-silvered thorn house in the evening, Nasibu emerged to lie at a distance, waiting for Nani and Nafasi to do something. Finally one morning Nani was nearing the end of her love affair and sat up as Nasibu plodded empty and hungry over the rise. But Nani was not looking at Nasibu. Beyond him a distant movement had caught her eye. A cheetah was chasing a gazelle, the long-legged cat a blur of motion as it gained on the fleeing tommy. The gazelle swerved and doubled back, dashing across the scrubby vegetation, followed closely by the lithe cheetah who steadily caught up. A little cloud of dust rose where the cheetah leapt and rolled over with the gazelle in her jaws and dew-claws.

Nani was on her feet and running up the slope before sleepy Nafasi was fully aware of what was happening. Nasibu changed direction and rushed to join Nani. The cheetah had only just opened the gazelle carcass when she saw Nani running at her. Immediately she gave up her hard-won meal to the much bigger lion. Built for speed, the cheetah lacked the strength either to carry her booty away or to defend it against marauding lions and hyenas. She took her disappointment with an appearance of resigned dignity and walked off to find a spot where she could rest before her next hunt.

Nani got a few bites out of the small carcass before Nafasi and Nasibu arrived to pull it apart. Having two large males as companions was not ideal, but the following night Nani was grateful for their company as they defended a kill from another group of nomads. Hunting, scrounging and scavenging, the trio stayed together throughout the wet season. The tan wildebeest calves were turning dark by the time the herds began to leave the drying plains. Instead of following them into the woodlands, Nafasi, Nasibu and Nani stayed on the grasslands, ranging widely, becoming known if not accepted neighbours of the Sametu pride. The trio began to roar their whereabouts to one another and to the other resident lions. They had taken a place on the plains.

Cheetah

CHAPTER SEVEN

Hard Times
(Dry season, 3rd Year)

An early dry season crept over the plains. It stalked the storms which turned tail and fled before the wind that sprang with force from the east. Having captured the plains for a long season, the wind panted over it daily, its hot breath steadily drying out the grass. It slowly blew away the scattered particles of prey that still littered the grasslands. The gnus retreated in long columns, trailing hazy ribbons of dust behind them; the other migrants left in a ragged exodus of herds and families. The tracks of a million departing animals were slowly erased as the fine soil was blown from the drying waterholes, mud flats and winding exposed trails. The dry season settled in, stretching its tawny form across the withering grass.

The Sametu pride was still living well, catching the departing migrants, when the first of a series of ominous events occurred. Salama vanished. The others accepted her absence without much notice. Salama had always been somewhat of a loner, gone for days at a time, and her loss was no hardship for the big pride. Her hunting skills had not been exceptional, and there would now be one less mouth to feed. But Salama left her three cubs behind when she met her doom somewhere out on the plains, from disease or wounds. Luckily for her cubs, they were the same age as those of Siku and Sukari. These two mothers adopted the orphans, giving them a share of their care, attention and milk. However, nine small cubs were a lot for the two to feed, especially as the dry season advanced and prey became scarce.

The Sametu pride began to make excursions westwards over Ratel Ridge into the broad Seronera watershed with its greater hunting opportunities. Sometimes the nine cubs would join the pride for its long treks, but often when the lions returned to Sametu for water or rest, the exhausted young cubs remained there, hiding together in the marsh or kopjes as the others left again to hunt. Days might pass before the adults returned. The young cubs grew steadily thinner, more wary and firmly united. Their isolation seemed to make them especially cohesive; they huddled or ran away together when any danger threatened and did not come readily when called. These traits made them hard to lead on hunts, so Siku and Sukari increasingly often left the cubs at Sametu.

As their milk dried up and their companions hunted further westward, it became imperative that the nine young cubs should keep up with their guardians. Siku and Sukari expended much time and energy commuting to

Black rhino and calf

Sametu and their presence, plus the food they provided, encouraged Kesho and Kali to hang about the area too, contributing to their burdens.

On the day they finally managed to get the cubs away, an amusing incident happened. The nine young cubs had been hiding in the marsh by the muddy puddle that was all that remained of the lake. They had been alone for days except for the occasional appearance of Kesho and Kali. The two males came and went from Sametu, often patrolling to the south, chasing away the encroaching nomad trio. They were away the evening the cubs saw the local rhino and her calf ambling down the valley. As she neared their hiding place, the cubs ran, bunched together like a pack of mongooses, furtively keeping to the thick vegetation that carpeted the valley bottom, scaring several marsh owls which frightened the cubs even more.

The rhino never gave any sign that she had seen the cubs but went to wallow in the muddy ooze near the faltering spring. She and her half-grown calf stayed around that night but had wandered up-slope in the morning when Siku and Sukari came back to fetch the cubs. The tired mothers checked the kopjes, the marsh and wells, calling and calling. When no cubs came forth, the females started tracking them by scent and finally found them far along the valley past the sentinel acacia. The cubs emerged from their bush clump reluctantly and slowly followed Siku and Sukari all the way back to Fig Tree Kopje.

Siku led the way up the rock, stopping to sniff at a huge ball of dung deposited precariously at the edge of the platform. The dung ball was stale and dry after days of sun and wind. The elephant who had deposited it had been wandering with a companion far from its usual wooded home and had found Sametu a convenient refreshment stop. The elephants had stripped the bushes and splintered the row of baby fig trees on top of the kopje. The mother fig was spared the fate of the young trees for she twisted up over the highest rock, which even the surprisingly nimble elephants couldn't climb. But the dry season had come to strip her of her crown and she looked down leaflessly on her shattered offspring.

Elephants at Sametu

Lacking shade, Siku and Sukari stretched out on the open rock in the wind while cubs fought over their dry nipples or pushed the great dung balls around the kopje top. Thus they were all safely out of reach but in an excellent position to view the return of the rhinos. Moving ponderously down the slope in the midmorning heat the two rhinos reached the far end of the lake and nibbled the

herbage around the lone acacia. Their pointed upper lips nipped off leaves and stems with a delicacy belied by their great bulk. Before long, their already slow movements slowed to a stop. They found sandy places and reclined to sleep, the mother with her head up and her calf on its side. They looked like typical Sametu rocks, except for the occasional flicked ear and the activity of the tickbirds which searched over the cracked hides for parasites. The rhinos relied on these drab birds with bright red beaks to warn them of danger. The tickbirds were their personal alarm system.

And the system worked. The birds flew up, wheeling and screeching, as Kesho and Kali passed by, returning from their long, hot march, eager for water and rest. The lions stopped in their tracks as the big rhino rolled to her feet with a sharp puffing snort. She stood blinking, swinging her head from side to side, trying to get the threat into her sight. With her head high, she advanced through the thick grass and up the bank, her calf following. Finally she focused on the two lions who stood quite still, staring fixedly at the two horned beasts.

The calm was shattered. Puffing and snorting, the rhinos charged. They chugged along with surprising speed, tails up, straight at Kesho and Kali. The two males turned and sprinted away as fast as they could go. Up the slope to Owl Rocks they ran, the rhinos pounding along behind. The owls flew out as the two males rushed in, climbing rapidly into the centremost and highest boulders. The mother rhino came to an abrupt stop halfway to the rocks as her would-be victims vanished. She snorted some intimidating remarks in rhino language and swung her head round, looking for the lions. The tick birds settled down on her mud-caked shoulders and her calf peeped round from behind her substantial rear. Neither seeing nor smelling any more lions, the rhinos returned to the end of the marsh. The owls flew back to their tree and Kesho and Kali emerged to sit on top of the kopje as though nothing at all frightening had happened. Even so, they did not come down.

The audience on top of Fig Tree Kopje had watched this event with undivided attention but as calm returned they stretched out to rest. In the afternoon, Siku and Sukari led the nine young cubs hastily up the slope of Ratel Ridge, seen by neither rhinos nor hungry adult males.

122

The two mothers had solved only one of their problems by getting the cubs to follow them. They still needed to find prey and also their fellow hunters. But that night, they did not find either. The following evening they scanned for food or friends from the side of Bustard Hill but saw only a pair of side-striped jackals busily digging. Running to investigate, the lions found a scattered array of round earthen balls, broken open. The shy jackals watched from a safe distance as the lions nosed the dung ball cradles. The broken balls had yielded their dung beetle grubs to the hungry jackals. Perhaps the grubs were some of the last pieces in the circular chain of life and death, for the balls had been buried in the previous wet season on the spot where the Sametu lions had shared two gnus with sundry other living things. Now that buried treasure was being unearthed by the jackals, an indirect gift from gnu and beetle and maybe even the lions, who left the jackals to their work.

Moving down into Warthog Valley, Siku, Sukari and the cubs saw some other lions. The two groups watched each other carefully before coming closer. The cubs were the first to abandon caution for they recognized Sam, Su and Si who bounded away to greet the newcomers. Sonara and Swala came next, then

Side-striped jackal and dung-beetle balls

Safi and Shiba. With enormous enthusiasm the lions milled around. It had been a long time since they had been together and they revelled in each others' company. Sam, Su and Si were tireless players and soon exhausted their younger relatives. At fifteen months, Sam, Su and Si were beginning to shed their milk teeth and were better fed than their five-month-old friends. But they were still learning to hunt and had much to learn about their environment. A lesson for the cubs was shuffling along nearby at that very moment.

Sonara was the first to see the porcupines. She left the group at a lope and circled the two carefully. Spreading their full array of quills, the two porcupines rattled their formidable armament in warning. The older lions inspected them warily, then continued down the valley. One porcupine lurched away and rapidly disappeared down a big hole. Sonara turned back and dug around the edges of the hole, sniffed about, then ran to join the others, most of the younger cubs trailing her.

Sam, Su, Si and two of the young cubs stayed to watch the other porcupine,

following it closely as it waddled along, huffing and puffing but not quite managing to blow the curious little lions away. Sam approached very close and sniffed at the spiny creature. The porcupine didn't appreciate this scrutiny and twisted around, spreading wide its thick body quills. It shivered its tail so that the hollow quills hissed together, stamped its feet and jerked backwards. Sam pulled back too late, receiving two fat quills in his inquisitive chin.

Sam grunted and fled. Si, Su and the other cubs ran too, frightened by Sam's sudden departure even more than by the porcupine. They rushed to catch up with the wiser lions who were making their way steadily down Warthog Valley. Sam kept shaking his head as the pride walked along in the darkness but he could not dislodge the quills. Stars shone brightly here and there through scattered clouds, and the air faintly tingled with the breath of an approaching thunder shower. It hadn't rained for many days and the smell of the storm was enough to invigorate the lions. They walked along with an almost jaunty air.

Late in the night the lions came to the edge of the big marsh and coalesced on a bank. Several of the females lay on a termite mound which rose from the bank and gave them a good view over the marsh. An old moon was rising in the east to glow dully over the dark sea of rushes, adding a lustre to the lions and the clouds above them. But about thirty large lumps remained stubbornly black — they were buffaloes, grazing quietly in the marsh. Keen hunger and a sense of daring impelled the hunters to risk stalking these big beasts. Sonara, Shiba and Safi crept down the termite mound and soon disappeared into the grassy space between the lions and the marsh. Swala and Sukari waited awhile, then they too slithered off into the grass. The cubs stayed on the bank with Siku.

The moon slid behind the approaching storm cloud that began to rumble as it sped over the plains. A few drops of rain fell, just enough to bead the lion's fur and pock-mark the dusty ground. Si approached Sam to lick drops from his head and neck, for the lions were thirsty as well as hungry. Sam was still twisting his head about, catching the quills on the grass and his paws, each time plunging them in deeper. Porcupine quills are not barbed, so can work themselves out eventually; meanwhile, however, the rewards for his inept bravery hurt. Sam lifted his chin as Si licked him. She sniffed a quill and caught it in her teeth; Sam moved his head and the quill came out. The other was imbedded more deeply but by wriggling his head Sam somehow caught it between his paws and it too was dislodged. Si sniffed the blood-tipped quill and began to chew on it, getting poked in the mouth by its sharp end. Next time the cubs met a porcupine they would be more careful.

A long time passed, more rain fell, then the storm moved on. There was a roll of thunder, but not from the departing storm. It was the many hooves of the buffaloes as they stampeded through the marsh. The stalkers had finally begun the chase. The cubs and Siku sat up and stared at the mass, running and bucking through the rush clumps. Safi was clinging to the back of a great black steed that jumped and twisted, trying to shake her off. He bellowed and snorted. Sonara leapt at his shoulder as he passed, her claws raking the tough skin but sliding off. The buffalo kicked her hard as he passed and she fell sideways on to the ground, winded. Safi rode on, slowly slipping backwards as her mount

bucked and bellowed. Finally she fell, rolling, a hoof slicing her as the buffalo
pivoted around, hitting her side with a massive horn. He would have killed her
easily had Sonara not suddenly stood up, recovering her breath and looking
around, enough to make the bull turn and run.

The herd of buffaloes reached the edge of the marsh and thundered up on to
the grassy flat and safety, forming a tight clump. They turned to stare back at
their attackers, lifting their heads high, trying to catch their scent and fix their
positions. Two big bulls emerged from the throng and walked boldly to the edge
of the marsh. The lions flattened themselves to the ground and began to move

slowly away. There was no hope that the buffaloes would provide them with a meal and they did not wish to be trampled should the herd decide to pursue them. The bulls were joined by others and the former prey became the hunters as they entered the rushes, looking for the lions. Creeping away as fast as they could, Swala, Sukari and Shiba reached the far side of the marsh. Keeping low, they came back to join Siku and the cubs. The lions watched the buffaloes move along the edge of the marsh, tense and eager to combat their foes.

Finally Sonara limped out of the marsh to join the waiting crowd on the bank. Only Safi was left in the danger zone but the others could not wait for her. At Sonara's movement, the buffaloes saw their victims and charged. The lions fled up the slope and away, leaving their antagonists far behind. The pride kept on going. Skirting the marsh they met the morning on the slope of the second hill, tired and hungry. Safi didn't find the group until the following night. She looked weary and her wound was oozing and raw. Many prey animals tried to impale, kick or trample attacking lions, but few were as deft or deadly as the buffaloes. They were also one of the few animals that banded together to protect themselves and the lions had learned that buffaloes in groups were unbeatable.

The females hunted their more usual prey after that; topis, kongonis, warthogs and gazelles. The lions were still learning about what was catchable and edible and were willing to try for almost anything as they grew hungrier and thinner. Sonara even stalked black-bellied bustards, but they, like most birds, escaped in a whirl of wings and dust. Lions just are not fast enough to utilize such creatures as food. The pride hunted far and wide over the hills and valleys of the Seronera basin. The three large and nine small cubs stayed together and dutifully followed their mothers, who went warily, seldom roaring, constantly on guard against the Loliondo males, other prides and even nomads. Kesho and Kali were loath to join the wandering pride and tended to stay nearer Sametu, protecting that relatively empty area and subsisting mostly on scavenged kills.

Then another member disappeared, but only for a time. Sarabi had been coming and going from the pride in her usual haphazard manner when at last her time to bear cubs had come. The pride was then in the farflung reaches of the outer Seronera valley where there were no good den sites. Sarabi had been forced to choose a hollow in the valley below the rounded hills, the only cover a clump of scrubby bushes. There was no water nearby, just mud patches when it rained. It was an ill-chosen den.

Sarabi was confined to her hollow while the rest of the pride wandered on. At dawn they caught a topi but their growls attracted two young nomad males and the females ran with the cubs, not risking a fight though the nomads were just youngsters and did not challenge the right of the females to their kill. The females and cubs had had enough to eat to make them thirsty but water was a long trek in any direction. They sat together on an open slope in the scorching sun with the wind drying their panting tongues, then turned at last to straggle up the long rise to Ratel Ridge. The torpid train of lions moved through the heat and dry grass with heads low, eyes half closed against the glare and wind.

Nasibu (left) and Nafasi. Males of the open plains usually have blond manes, which provide better camouflage than dark manes.

The Sametu pride relaxing on one of their kopjes.

Nasibu (left) and Nafasi by the Sametu marsh.

The authors camping at Sametu; two Loliondo males observe them from the rock.

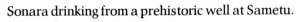

Sonara drinking from a prehistoric well at Sametu.

Evening on the plains.

Swala with small cubs.

Nafasi, Nasibu, and Nani — the nomad trio — on the plains south of Sametu.

Nafasi mating with Sarabi, soon after his acceptance by the Sametu pride.

Sisi, aged five. She was the first daughter that the Sametu pride successfully reared to maturity.

Nafasi asleep. This is how a lion spends most of its life.

Two females have just caught a warthog. While they kill it, a third hurries to join them.

In a few minutes, the hungry lions tear the warthog to pieces.

Left: Lengai feeds on a wildebeest stolen from the Sametu pride. Males can hunt their own prey, but they scavenge whenever possible.

Overleaf: In the evening, the Sametu pride wake up and start to scan the plains for prey.

The Sametu pride killing a topi. This large antelope was an important source of food for them in the dry season.

Sonara even stalked black-bellied bustards

Suddenly breaking into a run, Sonara galloped to a muddy patch with a little water and lapped as fast as she could. The others swarmed around the paw-sized patch of water and fought for places just as they would over a kill. In seconds, the water was completely gone and the few lions to get a moist mouth sat with dark mud dribbling down their white chins.

The group lay about the damp place, panting, tired and hot and still very thirsty. All around, the dry air shimmered and danced with far more energy than the lethargic lions. Vast mirages lay on the plain, rippling in the hot wind like true lakes. Sisi lay as though drugged, too hot to sleep, looking back towards the four hills squatting above long lakes of make-believe water. Distant plumes of smoke from the incessant grass fires hazed the far horizon. Sisi stiffened and stared as a mysterious form wavered into view. The moving mirage slowly materialized into a lion shape, shimmering closer and closer, gaining substance until it became Sarabi. Sisi rose to greet her. Sarabi wasn't her normal gay self and returned Sisi's headrub listlessly. Si sniffed all along Sarabi's bony side to make sure it was really her old friend, for Sarabi smelled strange, unhealthy, her coat was rough and dry and ribs pleated her thin sides.

Sisi followed Sarabi as the lonely female went to greet each companion in turn. Sam and Su and the young cubs all received at least a touch, then Sarabi went to the dampened spot and sniffed around, hoping for water. Finding not a drop, she lay down heavily on the damp earth. She had heard the growls of the pride and had run to join them, only to discover the two young nomad males hanging about, so she had tracked her family to find deeply-desired companionship. It had been many days since she was last with them and her condition had deteriorated noticeably.

Sarabi twisted her head round, trying to lick her unkempt fur, but she couldn't reach the thick clusters of ticks that encrusted her from head to tail. The hollow she had chosen as a den had been often used by other animals and a bounty of tiny ticks lay in wait for another suitable host. They found Sarabi

Sarabi's ticks

most suitable and buried their minute heads deep into her skin as she lay confined to the den. Without her companions to help groom her, Sarabi had been able to get at only a few of the thousands of baby ticks with her tongue. Many had latched on firmly, swelling to the size of grapes with the blood they needed to produce their own young. The dry season – lack of food and water – the isolation, long treks and her pregnancy had made Sarabi very vulnerable. Her strength was low, her milk meagre; without her hunting partners she could not get enough food to survive. The poor cubs that had precipitated this bad state were lying covered with ticks themselves back at the hollow. None of them would ever leave it alive.

Fringe-eared oryx

In the afternoon, when the sun lay heavily on the lions, Sarabi awoke to another portent. Over the crest of Ratel Ridge came several elegant antelopes, the likes of which neither she nor the rest of the startled lions had ever seen. Heralding the driest months of the dry season, the oryx moved quickly along the slope. They walked with superb grace, like princes, wearing their straight horns as though strapped to their heads by dark face-bands. The oryx came from the east where the land shelves its way down the great rift between the ancient and recent mountains. Seldom did they leave their arid haven; that they had come so far west was a sign that even these desert lovers had had to seek better pastures.

The lions never had time to consider how those swordlike horns could skewer them, for the oryx soon passed out of sight. Sarabi rose stiffly, stretching her weary self before returning to the fateful den and her sluggish cubs, their lifeblood slowly being sucked away. The others rose early too, heading for Sametu and water, meeting Kesho and Kali on the crest of Ratel Ridge. The pride went home together, Sarabi joining after a few more days, ready to start all over again, choosing Kali as her mating partner.

Sega had a litter next, at Sametu, a better den site but no better time of year. Soon, Swala and Safi had cubs too. The dry season lingered on and the lions grew thinner. Then came a disaster which was to have repercussions for many more months. Kali vanished. Probably succumbing to wounds received in yet another fight with intruders, he simply ceased to exist for the pride. Kesho roared and roared alone and finally accepted his solitary role. Without Kali, however, he seemed to become more timid, his patrolling was curtailed, he kept closer to the females and cubs, grew increasingly wary and cautious. The pride territory was no longer kept so well and boundaries grew fuzzy, more permeable to other lions.

The end of the dry season came after a remarkable celestial event, a full eclipse of the sun. The lions hardly noticed this strange phenomenon, that

Susu

occurred shortly after sunrise. The rising sun dimmed and the light on the plains turned from golden to brown, shadows softening as if in moonlight. Slowly moon and sun met and moved apart, a brilliant crescent of the sun blazing over the rim of the moon at the parched landscape, insistently reclaiming its day. Only humans can make such things into portents, taking the eclipse to symbolise the coming effacement, but eventual rebirth, of the Sametu pride. The lions merely lay in the weird twilight, and then suffered through the hot, dry day that followed.

Finally the dry season crept off the plains before an onslaught of riotous storms. Loudly proclaiming victory with thunder and lightning, massed clouds drenched the plains, relieving the long drought. The scruffy, emaciated lions welcomed the storms and the subsequent return of their prey. Eight of the adult females had survived and so had hardy Sam, Susu and Sisi, who were coming into their sub-adult strength. Five of the original flock of nine had endured through those hard times but only three of the cubs remained of the litters born at the worst time of the year. And Kesho remained alone.

Sisi

Sam

CHAPTER EIGHT

Waterhole Ridge
(*Wet season, 4th Year*)

Shiba left her new litter of cubs curled in the cleft and climbed to the top of Owl Rocks. Shadows from the rocks stretched out comfortably over the grass, sleeping long in the late afternoon sun. Vast, luminous clouds floated overhead, and the eastern sky was slaty dark with the promise of rain. It was the middle of the rainy season, the plains a deep rich green, the grass around the kopjes long and in full flower. Shiba settled on to her tummy on the highest rock and scanned around Sametu. She saw a few gazelles, zebras, topis and kongonis scattered around the valley. Prey was plentiful again and had been for the three and a half months of her pregnancy. Shiba and her cubs were healthy, the time of year was good and the den site one of the safest in Sametu.

Yet Shiba was vaguely unsettled. She longed to be with the rest of the pride. No other litters had been born at the same time as hers and she had to care for her brood without company. Solitary lion litters never seemed to thrive. Shiba and other lone mothers were drawn away from their cubs to hunt, eat and travel with their companions, thus the cubs would be left for long periods, alone and vulnerable. When litters were born at about the same time, the different mothers could form a mutual aid and protection agency, greatly enhancing the chances of some of the cubs surviving. So despite the advantages of the season and den site, the signs for the future of Shiba's new cubs were not particularly auspicious.

A shaft of sunlight pierced the froth of clouds and illuminated Shiba on her rock. Flashing brilliantly into the spotlight, an orange and black bishop bird flew past. He was on his way to pluck yet another grass leaf to weave into his globular nest. Soon he whizzed back past Shiba with the bright green blade in his beak. His mate perched on a hibiscus bush at the base of the kopje, watching him weave the grass into his nest. He had started three different nests, none of which was satisfying. This one was to be perfect, a work of art, an irresistible temptation to the female. Off he flew again, a dazzling speck, stray ember of the setting sun.

Eastwards a rain shower drew a curtain across the leaden sky and a glorious double rainbow straddled the horizon in a full arch. Softly, from beneath the rainbow, came the first roars of the evening. Shiba turned to listen to the welcome sounds of her sisters. Kesho added his booms to the chorus and the many voices purred out a great song into the twilight. The pride was often on the eastern ridge now. They had discovered that the ridge had a chain of waterholes where they could catch unwary prey coming to drink. From the elevated shelf above the well-spaced pools they could look back westward into Sametu Valley and also eastward into a shallow wide valley where many zebras and other animals grazed.

All the cubs except Shiba's new litter could travel with the pride, so there was no real need for the others to stay around the Sametu kopjes and marsh. As the migratory prey moved eastwards so did the lions. Many lion prides shifted eastwards annually during the wet season, stopping when their path was blocked by another territorial pride, or by the scarcity of prey. This wet season the Sametu pride had been partly tempted and partly driven to go further east than ever, extending their boundaries to the ridge of waterholes and part of the valley beyond. No prides claimed that area though the nomad trio did use it too. Hunting was good on Waterhole Ridge and the Sametu lions stayed there happily, utilizing its resources. It was unfortunate, as the pride would learn later, that their residency on Waterhole Ridge left the heart of their homeland unprotected.

Shiba felt very alone in the empty heart of Sametu. She groaned low after the last of her pride's roars dimmed along with the light. As if she had called to them, her cubs miaowed from their nook below in the rocks. Shiba grunted and half-rose, intending to go back to her den. A movement in the marsh momentarily distracted her. It was only the serval cat, leaping high after a rat, his movement attracting the attention of a host of small birds that twittered at the predator. The long-legged serval had become more open in his hunting around the Sametu area. Since the lions so seldom came to drink from the wells, he had begun to spend more time in the marsh, stalking the many mice and rats. The little birds were joined by big blacksmith plovers and their loud warning system ruined the serval's appetite. Unhurried, he moved through the marsh in his languid way, his spotty coat making him almost invisible in the gloom. Shiba lost sight and interest in the serval as another peal of roars rang out from the north-west.

These sounds made Shiba stand up quickly, stiff and uneasy. Closer this night than ever, the roars came from Bustard Hill; three Loliondo males and the three Boma pride females. Shiba stood tensely, staring north-west though the roaring lions were much too far away to be seen. The intruders had penetrated

Serval

the Sametu territory and no one would evict them. Kesho seldom chased trespassers any more, preferring to stay safely with the pride, roaring with the added voices of the females. He now began a reply to the encroaching lions from the crest of Waterhole Ridge and six female voices strengthened the retort. Shiba sat down to listen to the reassuring sound of the strong roars. The Sametu pride was vocal in its whereabouts and strength, however, it was obvious that they were in the eastern sector of the range, leaving Ratel Ridge and the Sametu Valley open. Shiba dared not roar alone and apprehensively went down into her den.

Clouds blotted out the weak starlight and a quiet, moist night blanketed the plains. A light rain began to fall, unnoticed by Shiba as she lay in the snug cleft, suckling her cubs. She dozed, but a loud groan from outside woke her from slumber. The low moaning was repeated, an incipient roar, coming nearer. Shiba rose and went out, recognizing Sonara's call. She was so glad to see the friendly shape materialize out of the darkness that she slipped on the wet rocks in her haste, having to leap on to the surprised Sonara who stood at the bottom. Rubbing and sliding affectionately on one another, Shiba and Sonara greeted, then settled into the damp grass at the base of the rocks. Shiba nipped at the long grass stems and chewed on the beaded wet blades. Sonara nibbled the grass too, but seemed tired and soon stretched out on her side. Suddenly she sat upright. Shiba had heard the sound too and both females went round to the western side of the kopje to listen more carefully.

A distant set of roars rolled from the west, so faint as to be barely heard. The roars evoked deep feelings in the two listeners. They were the voices of their mothers, aunts and cousins; the old females of the Masai pride. These had come out on to the plains with their large cubs to avoid some new males who had recently come to mate with the younger females of the Masai pride. This event was a reenactment of what had happened four years earlier, when the young females destined to become the Sametu pride had taken refuge on the plains with their mothers, running from the Loliondo males who had come to court.

137

No wonder the old Loliondo males were heard more often on the plains now, for young vigorous males had taken their place with the Masai pride. Little did Shiba and Sonara know that soon they and their sisters would be running once again; the presence of part of their old pride was a portent of things to come.

Sonara and Shiba answered their kin to the west, roaring loud and poignantly to their mothers and neighbours. A greeting, a warning, the roars spread out across the grassland to all the listening ears. The Sametu pride responded first, from the eastern ridge. Then the Loliondo males roared from Bustard Hill, letting it be known that they were still in charge of a huge kingdom, a force to be reckoned with by all prides around. Occupying the highest hill, they were heard even by the trio of nomads to the south-east. Nafasi led the reply, joined by Nasibu and Nani.

The sound of roars rested in silence for a time, then the high clear keening of golden jackals filled the air. Sonara and Shiba returned to the east side of the kopje, stopping to listen again. The jackals' song did not hold their attention, it was the sound of the lions' growls and hyena laughter. Sonara and Shiba trotted away from Owl Rocks going steadily up the slope of Waterhole Ridge. They found the Sametu pride massed around a dark carcass, hyenas humping around the perimeter, jackals darting in and out. The two newcomers sat down at the edge of the scrum to wait for a moment in which they could ease themselves unnoticed into the throng. Out of the dark came another latecomer, Kesho, who also sat down at a distance, waiting for his chance. Slowly Sonara and Shiba crept into the mob, then Kesho barged in too, deftly removing the almost bare spine and skull, carrying it off to eat alone in kingly splendour.

The rest of the lions snuffled about, looking for leftovers. Between them they had taken the largest share but were still hungry. Eight adult females, three sub-adults, five large cubs, and three small ones, plus Kesho, made a large number of mouths. The twenty lions were healthy and had eaten well for the last few months. Even so, they had to be constantly on the lookout for food. The females' increasing skill at hunting meant the lions were putting on weight. Certainly Su was fat, and Sam too, their stomachs huge, their tails thick. Sisi was still slim and muscular. She often hunted with the other adults and was now a fully accepted member of the team. Sam and Su were more opportunistic, seldom joining a hunt; however, their size and assertiveness gave them an advantage at kills.

At two years of age the sub-adults were lucky to remain with their pride. The adult females seldom made much effort to push their offspring out and Kesho did not seem to notice that his son was almost as big as he was and showed a definite, although small, mane. Of the three sub-adults, only Sisi contributed directly to the pride's welfare but all were playmates and guardians of the two sets of younger cubs.

Sam went over to one of the three small cubs and sat cleaning himself while swatting the little cub with his fat tail. The cub leapt at the tassel and chewed on it but Sam pulled his tuft away, slapping the cub again. The other two small cubs joined in, leaping on Sam. He batted them gently with his big paws but took care to restrain himself. The year-old cub that joined in the game next was likewise careful with the younger cubs but played hard and rough with Sam. The boisterous cubs rolled, wrestled and dashed about in between the older lions who lay scattered about, licking, cleaning, their fur wet from the light drizzle. Sam led a chase through the obstacle course of bodies and the resting females glowered and growled at him as he bumped into them.

Safi rose to lead the general trek away from the kill, up the slope towards the ridge top and a waterhole. Sam leapt at his mother as she passed by, but she twisted sideways, baring her teeth, then going on. Sam went over to sit on the peacefully unaware Sonara who was licking her paws. He slapped her with his tail; she grunted and rolled to shove him off, growling.

Away Sam raced, young cubs chasing along behind. The happy throng bounded up the rise, the adults listening carefully to distant roars, and roaring in reply from time to time. Sisi frequently added her voice to the chorus, another sign of her maturity. Sam and Su were hardly better at roaring than the younger cubs who miaowed and squeaked when they tried at all. Kesho remained behind, often roaring while he finished chewing on his share of bones.

Peaceful treks after a meal were voyages of discovery for the youngsters who played along merrily, exploring their domain. Two cubs saw a moving rock and pawed at it. The tortoise hissed and pulled itself into its beautifully mottled shell to resist the playful attacks. The little lions chewed on the tortoise's edges and sniffed at the wide cracks to no avail. Finally they left it and chased after the rest of the pride.

They caught up to find some of the younger cubs peering at a new attraction. It was a prickly ball, no bigger than the paw of one of the giants

circling around it. When a large cub picked it up in his teeth, the ball jerked and clucked softly, causing the startled lion to drop it. Two cubs stayed with it, batting it to and fro while the rest of the pride tramped past them. Then Sega came back and called the cubs to join her, ignoring their plaything. They followed reluctantly, and as their padded feet thudded out of earshot, the small hedgehog uncurled. Like a clockwork toy, it scurried away to hunt insects through the endless jungle of grass.

Reaching a flat, sandy place below the ridge crest, the pride elders stopped for a rest and to chew on the wet grass. They watched the dim forms of gazelles move ghost-like in the darkness on top of the ridge when suddenly some rather different shapes drew their attention. The cubs and several of the adult females clustered curiously around two inert balls. Two pangolins, or scaly anteaters, had been caught in the open with only their slow wits and armoured plates to protect them. The curious crowd of lions watched the coiled animals carefully. They were warm and smelled like food but how could they be opened? The pangolins wore imbricated plates that were impervious to lion paws. The scaly beasts were too tough to bite into, too big for lion jaws to grip. Sarabi pushed

Hedgehog

one with her paw, gently. Nothing changed. Sonara and Si pushed together and the pangolin rolled over. By keeping its tail curved tightly over its vulnerable body and head, the pangolin presented an impenetrable surface that had no corners, rolling about easily as the lions pushed and shoved. After much sniffing, poking and attempts to gnaw, the adult lions grew bored and began to lie down again. But the three sets of cubs remained interested and stayed by the two plated balls, panting over them in their efforts to produce some response.

The pangolins pulled themselves tightly together whenever a lion touched them, but they did not uncoil when left alone. Finally Su managed to get her huge maw open wide enough to get a grip on a pangolin's tail. She picked it up and hoisted it aloft. The pangolin dangled, partially uncoiling as its muscles strained against gravity. Su lowered it to the ground, grasping it with her claws, trying to get her teeth into the hard armour-plated hide. Sam pushed it with his paw and Si smelled it all over. The adult lions began gradually to drift away and at last the cubs ran to catch up with their mothers.

When the lions were long gone, the pangolins uncurled warily. They waited in the grass, finally feeling safe enough to unfold, plate by plate. At last they

140

stood up firmly on stubby legs, sniffed around and peered into the darkness. The two strange creatures shuffled away from the scene of their torment, continuing their interrupted search for termites. Like tortoises, hedgehogs and porcupines, what they lacked in speed or brains they made up for by being endowed with suits of armour and determination in the face of danger.

Dawn began to pull the darkness apart, leaving the sky full of tattered clouds that caught the eastern glow. The lions reached the crest of Waterhole Ridge.

Pangolins

Stone-curlew and three-banded plover

Poised on top of the bank, Sonara looked down at the nearby waterhole. A narrow sandy beach curved on the near side and a rubble of stones and muddy clods rimmed the other where the slope led down into the sweep of zebra valley. The pool lay still and pristine in the early light. Stone-curlews became noticeable when they moved, searching for hidden things. When they stood still and tall, they vanished again, so well did their colouring match the background. More conspicuous, pretty three-banded plovers scurried around the edge of the pool, jabbing their beaks into mud and water.

The cubs were still exploring. Some of the larger ones had spread out to dig and sniff at the entrances to spring hare burrows which pocked the bank. The occupants were deep in their warren beneath the green sward, taking refuge from both predators and daylight. They grazed only during darkness and kept to the safety of their holes during the dangerous hours. Lions prowled from dusk to dawn so were always a threat, but no paw could reach the spring hares underground. The cubs peered into the many large and small holes that riddled the shelf above the nearby waterhole. Other animals used the holes besides spring hares and the place was full of interesting smells where mongooses, jackals and foxes had lain. The rumps of investigating cubs presented irresistible targets for their playmates who sneaked up on them, trying to hide their approach behind the small grass clumps. Pounce and pursuit followed as the cubs rollicked about in the growing light of morning.

142

Su led a contingent of cubs down the bank to the water's edge where a pair of the inevitable blacksmith plovers protested with loud clinks. The bank, sandy shore, holes and tussocks all offered the cubs endless opportunities for play. Soon they were racing across the shore and up the bank, only to fall down again. Sam barrelled along after a fleeing cub who zigzagged through the maze of spring hare holes and jumped off the bank. Sam careered over too, flying straight into Su who was clambering up. She slipped backwards sitting heavily on two little cubs behind her, for Su was very fat indeed. The hardy cubs inhaled gratefully as Su rolled off them, wrestling with Si who had joined the brawl. The small cubs leapt on to their two bigger playmates, attacking Su's thick neck and chubby face, embedding their teeth in the rolls of loose, flabby skin.

The sun peeped over the eastern horizon making the grass and water glitter, caressing the lions with a golden glow. Sonara turned her head back to Sametu when she heard Kesho's roars. She composed herself for a reply and joined in with the others who also answered Kesho with a rolling rhythm. The last series of soft grunts fell into the limpid air while the sun rose.

Sam went along the ridge to watch Kesho plodding forward, coming up the slope to join the pride. Shyly he approached the big male and tried to nuzzle him in greeting. Kesho ignored the solid-looking sub-adult and went on, Sam following him. Kesho stopped to accept Sonara's greeting. She slid under his chin then gave him a playful whack on the nose. Kesho sniffed her carefully, following her a short way as she flicked her tail in his face. He then went to a grass clump to scrape, urinating on his feet and tail as well. Sam took the opportunity to sneak up behind Kesho and sniff his powerfully smelling genitals and tail. He wrinkled his nose in classic lion style as he inhaled deeply. A middle-sized and a small cub joined Sam for a good sniff at Kesho's marked spot after the latter left to go along the shelf towards the waterhole.

Kesho stopped on the bank overlooking the pool and his three admirers assembled next to him. The four males of different ages formed an interesting tableau as they stood on the bank, backlit by the rising sun. Kesho was almost eight years old, nearing the end of his prime. Sam was just over two years, needing another two years of growth, a full mane and much experience before he could be considered a fully adult male. Next to him sat a half-grown cub, just over one year of age, looking like a slightly bigger version of the seven-month old cub standing next to him, though his crest of hair was more noticeable.

Kesho went over the bank to drink at the waterhole with his train of followers. They watched with awe as he bent his elbows and drank awkwardly, rump sticking up into the air. Sam put his face close to Kesho's genitals and took another good whiff, then pulled back fast as the older male lay down, continuing to lap the water and swishing his tail in warning. Sam and the two cubs waited until Kesho had finished, then followed him back up the bank. Kesho walked sedately across the grass to sit alone on his green throne overlooking the ridge and two valleys. The three young males sat at a distance but were distracted from their idol by an invasion of playful cubs led by Sisi and Susu. Off they dashed, leaping and tumbling, expending their energy before the

heat of the day slowed and finally stopped them.

Clouds were soon marching across the sky in broad ranks. The air grew oppressive and the lions sought rest on the damp sand and mud by the pool. They panted as they lay in the humid heat. Even the quick gazelles grew less active, posing on the skyline like carved statuettes. Kesho joined the lions who edged the waterhole, taking another long drink and carefully sniffing Sonara again. She would be ready to mate soon. As sole consort, Kesho had to be very active should any female desire his favours. Luckily for him, most of the females were in the long quiescent period that went with the growth of cubs. Most females did not mate again until the cubs they had suckled reached the age of a year and a half or even older. Sonara was the last of the mothers of the first litters that had come into heat. She had mated with Kesho a few times already but had not become pregnant. He was duty bound as chief and only adult male to try again.

The twenty lions lay wearily around the waterhole. As the day oozed along, they sought higher places on the upper shelf where just a slight breeze fanned them. Si sat among the others for a while as heads dropped and all stretched out to sleep. Her curiosity aroused, she crept up carefully on yet another interesting creature that occasionally shared her world. The creature was resting in the midday heat, flaps open at its sides to catch the breeze. Si slowly moved around the silent smelly creature, creeping closer to a strange pale object lying half in and out of the shade under its legs. She went cautiously – from past encounters she knew that the big beast sometimes moved abruptly. Finally she was close enough to risk a sniff at the sprawling shape. Rearing back in alarm when the shape suddenly flung itself upright she dashed to safety and looked back to see the small creature swallowed up by a slamming flap in the large angular object which had sheltered it. Si lost interest, the big "animal" was immune to attack,

144

like the rhinos, tortoises, pangolins and so many others. She left the car and its human occupants and returned to the pride, taking up a sphinx-like pose, then falling asleep among the others.

On the shelf, lions lay scattered on their sides or backs, white bellies up to reflect the sun's fierce stare. Only Kesho and a middle-sized cub had their heads up as the midday heat pressed down. Kesho was dozing, eyes mostly shut, but the cub saw a many-legged beast looming up. Shimmering along in the heat haze came six giraffes. The cub leapt to his feet and ran to hide over the brow of the bank. The rest of the lions awoke at once, the cub's fear communicating itself in a flash. They turned in unison to gaze at the giraffes as they strolled

magnificently along, moving like a disjointed animal, coming apart and joining up again.

Giraffes seldom crossed the plains and the young lions had never seen them so close. The cub who had first sighted them crawled up the bank and went over to lie beside Sisi, staring at the giraffes as they passed. The giraffes spotted the lowly lions and gazed down from their great height. The Sametu pride never hunted giraffes and had no idea of them as prey but the giraffes trusted no lion and kept them carefully in view as they glided away, leaving the lions to resume their rest.

In the late afternoon, raindrops began to fall and the lions stirred. They stretched, yawned, went to the pool to drink among the spreading circles from the raindrops, to solitary toilets or to join one another, greeting and licking. The cubs began to play again, sliding about in the sand and mud, rolling over in the grass or down the bank. It began to rain harder, a patter of raindrops followed by a torrent and finally a deluge. The lions became inactive once more, huddling against the heavy downpour. Sheets of water covered the ground, forming rivulets that began to flow into the pool, collecting in any hollow and seeping rapidly into the sandy soil.

The spring hare holes all over the bank began to fill and the floods forced their occupants out. The poor bedraggled spring hares were reluctant to appear above ground so early and with so many lions around, but drowning was the alternative. Si saw one emerge and hop away, its wet tail weighing it down. She chased it and it bounded along, zigzagging, eluding her. She gave up and turned back, almost stepping on another half-drowned spring hare just coming out of a hole. Deftly catching it, she went off to eat while the other cubs and a few of the adults joined in the scramble after the sudden exodus of these strong jumpers. Rain continued to pour over the plains, saturating everything. Sonara and

Spring hares

Shiba both caught slow spring hares and even one of the large cubs managed to get one. But most of the hares easily escaped and soon all were gone from the colony – a lesson in hunting for the younger lions and a tasty reward for the quick or lucky older ones.

All light was squeezed from the plains as night came down on the thick cover of clouds, wringing out more rain. Four females disappeared into the darkness to hunt bigger prey, going down the slope into Zebra Valley. The waiting crowd on the bank watched them vanish. Keen-eyed Sarabi turned her attention from the hunters to an interesting movement behind the flock of cubs. She saw a stray group of gnus with several young calves at their sides and sidled away followed by Sonara, Shiba and Sisi. The gnus blundered about in the rainy darkness. Smelling the lions, they panicked, running in all directions. Sarabi chased one and caught it deftly by the muzzle while Sisi hung onto its rump, and Sonara pulled it down. Lions began to appear out of the darkness to join in the feast. Opportunistic Su saw an orphaned gnu calf wandering alone in confusion. She waited until it was close enough, then with unusual swiftness she caught the calf and hung on to its neck inexpertly, waiting for its struggles to cease. Sam came over to see what she had caught and Su growled low at him. Picking up the calf she carried it off into the lion-free perimeter to eat by herself.

The rain continued to fall as the lions ate their meal, enough for everybody to have a full belly, especially since voracious Su had her own dinner alone, leaving enough for at least two or three other lions at the main dining area. Having stuffed himself, Sam began to chase hyenas for fun, then went to lie beside selfish Su, waiting to see if she would share the smallest morsel. Si joined him but neither got anything. Small cubs soon found their big playmates and enticed Sam and Si away into their games. The rain pattered on the twenty backs as the lions sat or played among the puddles on Waterhole Ridge.

Shiba finished cleaning her face and set out down the slope into Sametu to return to her cubs. Roars from the north made her stop to listen to the dreaded voices of the Loliondo males and Boma females. Behind her on the ridgetop, the Sametu pride began to chorus a response. Somewhat reassured, Shiba continued down the slope alone, back to the abandoned Sametu kopjes where her hungry cubs awaited her return.

The bishop-bird wife saw the lion pass but stayed sitting snugly on her eggs in her beautiful rainproof home. The serval cub saw Shiba climb up on to the rock from where he hunted on the near side of the marsh. The steinboks watched Shiba disappear into the cleft and the owls observed her presence in the morning when they returned to roost. They all got used to Shiba coming and going while the rest of the pride stayed away on Waterhole Ridge. The grass grew long and rank, the bishop-birds had babies to feed and Shiba was gone the day when the serval ran from the marsh in alarm. Several lions had appeared at Sametu and were sniffing around. But they were not the Sametu lions, come home for a visit. They were intruders.

CHAPTER NINE

Neighbours and Trespassers

(Onset dry season, 4th Year)

The Boma pride and their accompanying Loliondo males came to Sametu. Ripples from their arrival spread commotion up and down the valley. The three Boma females brought their cubs past Fig Tree and Jasmine Kopjes, crossed the muddy trickle that still flowed from the lake, and lounged beside the end of the marsh. Slinking away through the rushes at the other end of the lake, the serval cat startled several sleeping marsh owls who fluttered up out of the grass. They flew around the sentinel acacia and sailed down into the safety of the thick grass further along the valley bottom. A well-hidden caracal watched the mottled birds disappear into clumps, carefully noted their whereabouts and slowly began his stalk.

The Boma pride females scanned the kopjes, marsh and lake carefully from the bank while their three cubs stalked avocets and ducks. Startled flights of cape wigeons slapped away to the far end of the lake. Blacksmith plovers screamed and clinked at the intruders, swooping low over the heads of the marauding cubs, trying to distract them from their fluffy babies who hid motionless among clumps of grass.

From Fig Tree Kopje came the loud protests of the roosting pair of Egyptian geese. They scolded the big male lion who stood on the rock platform below. Lengai ignored the commotion. He looked over to the Boma females and cubs, then down to the base of the kopje where his brother Laibon was sniffing among the long grass. Opposite them, a third Loliondo male, Lerai, was exploring Jasmine Kopje. Lengai ducked as the furious Egyptian geese flapped away over his head. He then started sniffing carefully around the bases of the struggling row of young fig trees, the bushes, grass, and entire kopje top.

Caracal

The three Loliondo males sprayed and marked everything that could hold their scent. Lerai even sprayed a deserted bishop-bird's nest which was hidden among the yellow hibiscus and pink-flowered *Achyranthes* stems at the bottom of Jasmine Kopje. Taking their time, they staked their claim to that section of Sametu, incorporating it into their newly extended territory. The scent of the Sametu pride had been almost washed out by time and the long rains. Now the powerful new odour from the Loliondo males permeated the place, proclaiming new owners. When they had finished marking and exploring the two kopjes and adjoining areas, the males dragged an eland carcass into the fringe of the grass under the tall rock of the Fig Tree Kopje. They had helped to kill the big eland and had shared it with the Boma females and cubs. Now they began to gnaw on the bones at leisure, in the shade of the rock.

Lengai, Lerai and Laibon were scarred and worn, like their two absent brothers, Leo and Lemuta. They were all nearing ten years of age, a ripe age for male lions, few of whom survived that long. That the five Loliondo males still possessed prides and could expand their territory was living testimony to the advantages of numbers, of having age-mates and fraternal bonds. The five males had sired a prodigious number of cubs, had held up to four prides and a huge territory.

At the time the Sametu females first joined up with Kesho and Kali, the five Loliondo males were in their prime. They had been mating with the Boma females, guarded their rights to the Masai pride, kept up a waning relationship with the Seronera pride, and soon took over the Nyamara pride that Kesho and Kali had abandoned. But eventually some new males had appeared, three old ones at first, followed by their five young sons. Even

Marsh owl

152

Bateleur eagle mobbed by crowned plovers

now the Loliondo males could not compete with the influx. They could not continue to defend their vast territory and all the different prides from invasion, so had gradually withdrawn, moving further out on to the plains, relinquishing rights to the Seronera and Masai pride areas but extending their range east and south. The formidable five still had a future, and that future involved the Sametu pride.

The big males gnawed on their carcass and took turns napping. Lengai watched the three Boma females and their cubs wander up to Owl Rocks. A storm spread across the sky; the air was still and heavy. Steinboks' ears flickered from the throngs of tickling flies as they sank into the grass when passed by the strange lions. Like some advance guard of warriors exploring a new terrain, the females and large cubs spread out to investigate. The capped wheatear bobbed up and down on his rock, worried to see so many lions again, poking their noses into holes and bushes. The owls watched wide-eyed as the squad entered the kopje and began to pry into nooks and crannies. The prolonged sniffing and nosing about made the owls apprehensive. They were used to the solitary Shiba coming quietly to visit her cubs but this mass of lions combing the rocks bothered them. When a big cub pounced on an innocently sunning lizard under their tree, the pair flew off to find another roost, leaving the intruders to explore the deserted kopje, where Shiba's cubs lay huddled together in mortal danger.

Piercing shrieks awoke Shiba and the others of the Sametu pride who lay near a waterhole on the ridge to the east. A huge bateleur eagle swooped out of the sky and flapped away, chased by a squadron of screaming crowned plovers. The lions looked around but most did not bother to sit up. Shiba rolled on to her tummy and gazed at the calmly grazing gazelles that lined the ridge crest. She

153

stood, stretched, then climbed up the bank where she could look back down the slope into Sametu. In the valley bottom only the faint roundness of kopjes could be seen in the cloud shadows. Shiba sat down, yawning, and looked at the vague curve of Owl Rocks. She had left her cubs there early the previous night, having been forced away again by her hunger for food and companionship. Her cubs were still isolated, not able yet to travel with the pride. Shiba would return to them after she had managed to secure one good meal in the company of her sisters, but the lonely treks to suckle them were a duty she performed without eagerness. She lay down, putting her head on her paws, continuing to gaze down into Sametu, growing sleepier and finally closing her eyes.

The afternoon heat and light was slowly replaced by a sultry gloom as great grey clouds smudged the evening sky. Gradually the scattered members of the Sametu pride awoke, stretched, greeted and began to scan for prey. The lowering sun struck across the plains; its yellow light tinged the valley full of zebras. Two stallions were fighting while the mares and foals stood about restlessly, waiting to see the outcome. The thuds and thumps of the stallions' hooves came to the watchers' ears and hungry Shiba set off silently down the slope, followed by several other stalkers. The sun blinked as a black cloud struck the sinking eye, then it was sucked down into the murky, swirling clouds.

Rain splattered the ridgetop and smeared the view. The waiting lions lost

sight of the hunters. There was the sound of hooves running, then a long time passed. When the rain had eased off, Sega and Safi led the cubs along the ridge-crest. Kesho remained on the ridgetop, waiting for squeals and growls to announce a kill. A few stars gave enough light to show that the zebras had moved down to the floor of the valley; presumably their stalkers were having to follow. The lions on the ridge began to descend, then stopped to watch as several groups of zebras stampeded away, hooves loud on the soggy ground. Dim shapes of lions sprang mushroom-like from the zebras' wake – another unsuccessful hunt. The wary zebras fled, galloping away in their tight family units, the stallions last, ready to kick any lion hard in the jaws. Lions were less trouble to them than younger stallions who continually challenged the ownership of their harem.

Gradually the lions came together for a post-hunt reunion. They milled around in the darkness, then suddenly stopped still when a large lion loomed out of the night. The females and cubs stared, for the bearing of the heavily maned form was not that of Kesho. As the strange male strode forward, the cubs did not hesitate to turn and flee as though their lives depended on it. And, indeed, they did. The females turned and ran too, not quite sure who it was that walked so aggressively.

A whiff of the male followed them and some of the adult females turned round to face their pursuer. They had recognized the fearful odour of Lengai, their old enemy. Head held high, Lengai strutted forward. Several of the females edged more closely together while the cubs and other females fled. The opponents faced each other. They seemed to charge simultaneously, Lengai rumbling. The Sametu pride defenders swung their claws, raking the hide of the plunging Lengai. Snarls, slaps, growls and fur flew. Lengai fought back, swinging his great paws as the females leapt at him.

Another male lion loomed out of the dark behind Lengai. At the sight of Lerai the females turned and ran. Lengai and Lerai did not bother to chase them. They stood together, shoulder to shoulder, and began to roar mightily, the sound filling the dark night. Their challenging duet was answered by their brother over the ridge on the slopes of Sametu, and the distant Boma females roared added insults to the fleeing Sametu pride.

It was far into the night before the scattered Sametu pride members tried to find each other again in the now dangerous darkness. Greetings were prolonged. Each greeted the other with much sliding, rubbing and licking, reassuring one another, sharing and strengthening their collective scent. Near dawn, most had come together to rest on the floor of Zebra Valley. Kesho was not among them and Shiba and Safi were still missing, so were four cubs.

As light diffused again over the leaden sky, Shiba appeared briefly. She hurriedly greeted Sarabi at the edge of the group, then went up the slope to the ridge. Quickly moving along the top she disappeared over the crest, heading briskly back to Sametu. She was worried about her cubs and walked in haste to check on them.

The sun bubbled up, a huge blob of molten gold. The cluster of kopjes on the next ridge east struck a corner of the orb as it rose into the cool grey sky. Broken rays scattered across the valley and smote a walking lion. It was Safi returning to the group alone. Her coming was heralded by the rapid withdrawal of a hidden cheetah which spurted out of the grass as Safi came near. Three furry, ratel-like cubs scampered away after their mother. Safi paused to watch them run. Cheetahs were like nomadic lions, with no fixed abode, but temporary ranges. Was the Sametu pride to join the ranks of drifters? Safi walked on towards her kin. Sisi, Susu and Sam ran to meet her, greeting her joyously, the rest coming too; a grand reunion as the sun escaped the horizon and sailed brilliantly into the sky. Shreds of storms soon began to gather and as the sun rose higher into thickening clouds the lions arrayed themselves among the scrubby vegetation for their midday rest. They were tired, worn out from the trauma of the preceding night, and not refreshed by any meal. Sega and Siku woke early in the evening to set out in search of the missing cubs. The others slept on, then slowly began to ready themselves for the evening hunt, scanning the shallow curve of the valley, looking carefully up to the ridge with the waterholes.

Sega and Siku found the cubs far down the valley hiding together in a grass clump. Sonara met them coming back, nuzzling each cub and greeting her two companions thoroughly. Soon the entire group was walking eastwards as if they all had made a unanimous decision to avoid the ridge of waterholes for the time being. The company came to a halt when the first evening roars came from that direction, behind them from over the ridge towards Sametu. The three Loliondo males roared with the Boma females. Kesho did not answer from wherever he was; nor did the assembly of females and cubs who turned and moved rapidly eastwards. Darkness obscured them as they crossed Zebra Valley

and began to ascend the long, broken slope that led to the little group of kopjes.

The slope was cut into steps by wind, rain and the hooves of countless generations of animals trekking over the fine soil. Sturdy little herbs and grasses clung to the edges of the steps and a particularly spiny plant filled all the interspaces, prickly under the lions' paws. Hiding among the scrub and erosion steps, the lions managed to stalk and catch a zebra foal. They ate quickly and quietly, hungry and afraid. More roars hailed them from far across the valley, and strangers' voices came from the south-east. Finishing their small meal in record time, the Sametu pride continued eastwards into territory which they hoped was unclaimed. The lions passed the strange kopjes that glowed pink in the early morning light. Stopping to rest in a patch of prickly herbs, they stared at the rocks, catching a movement there. Sam and Sisi followed by Susu and some of the smaller cubs ventured closer and frightened away two young male cheetahs who had caught a gazelle near the edge of the kopje cluster. The startled cheetahs fled, perched briefly on top of the warm pink rocks shot through with crystal, then leapt to the ground on the other side and loped away in the graceful stride only cheetahs possess.

While Sam and Susu fought over the abandoned carcass, Sisi and the cubs explored the kopjes, then returned to the rest of the group lying in the open further up the short ridge. The pride did not rest easily all that humid day, but scanned the surrounding areas, sleeping in short bouts. In the evening they heard roars not only from the direction of their former home but also from the east, not far away; more roars came from the south just over a rise. Surrounded by unfriendly sounds, Sonara led the group west, back down the eroded scarp into Zebra Valley. She and Swala caught an old gnu who hardly tried to flee, and Sisi ended his life by a secure hold on the thin neck. Hungrier this night, the females and cubs growled more loudly over their dinner.

Suddenly out of the darkness came a lion trotting quickly towards the group. The diners froze, hunched over their meal. But the cubs recognised who it was and ran to meet Shiba. She passed them by with hardly a sign and ran straight to the carcass with none of the usual lion preliminaries. Shiba was ravenously hungry. It was the first meal she had had in days. When she had returned to Sametu, she had found the intruders occupying the kopjes. They would not let her come near the rocks, chasing her away repeatedly. One of the Loliondo males had almost caught her. Next night she had tried to approach again to no avail, and was chased long and hard. The trespassers drove her further onto the plains and she had run and run, over the ridge of water holes and down into Zebra Valley. Hearing the pride on the gnu kill she had run again, this time to food and to her friends as well as from the dangers behind her.

The scrawny carcass was soon bare and Susu and Sarabi growled low as they hung with claws on to the head and neck. The rest of the group sat about unsatisfied, cleaning themselves and wondering where they could go. Directly south came a loud trio of roars that made everyone sit up immediately. Rising quickly, the cubs and females started running. Again the roars sounded, closer this time and Susu and Sarabi left their debate over the bones to run too. The stream of Sametu lions flowed away down the valley as rain began to fall. In their wake came the three nomads, Nafasi and Nasibu in front, skinny Nani right behind. The males chased the fleeing Sametu lions for a way, then returned to sniff the spot where they had lain. Nani was more interested in the bones, glad that her two companions were preoccupied with smells. Nani was as thin as usual, perpetually hungry; it seemed that every time she caught something it was taken away or at best shared with Nafasi and Nasibu. The males were well fed, glowing with health and vigour, very interested in the flood of scent left by the Sametu females. They marked and roared loudly.

The Sametu females heard them and kept running across the wide valley floor towards Waterhole Ridge. Where were they to find rest and security again? Now it was raining and the plains still green, but soon the dry season would come again. Hunting in unfamiliar territory would mean that a difficult season could become a dangerous one. They paused in the middle of the valley to rest and listened to the nomad males. Nafasi and Nasibu roared mightily while silent Nani kept to her gnawing. Answering roars came from afar, over

158

the rise towards Sametu. It was not Kesho but the intruding Loliondo males. The females started to run again. Worn and weary they reached a waterhole by early morning. Kesho wandered nonchalantly into the group and was greeted by all. He accepted their affection stoically, even allowing Sam to sniff his mane and slide along his flank. Then he climbed up the bank to sit alone in majesty, a dispossessed monarch, reluctant to depart from his throne yet unwilling to defend it either.

Shiba left the others as they lay on the ridge crest. Apprehensively she headed down the slope alone, back to Owl Rocks. No lions were there when she came back to Sametu but many eyes watched her climb into the kopje. She found no cubs. Going over the entire kopje, crack by crevice she searched for the ill-fated litter and then explored the whole of Sametu but found only the reek of the intruding lions. Returning to Owl Rocks she rested in the midday heat. Her milk was drying up after the several days without suckling the cubs. Already her body was preparing to mate again. Shiba sighed and slept, totally exhausted. In the evening when the owls flew off to hunt, Shiba woke and began the long walk up the slope to rejoin the pride. The serval and steinboks watched her go; later that night they watched the Boma females return to the kopjes, sniffing about with proprietary interest, then lying in the place Shiba had vacated.

Weeks passed and the Sametu pride took to a wary, shifting existence, running, ranging far, avoiding strangers, intruders, neighbours. Kesho was the most careful of all, keeping close track of his enemies' whereabouts. He was often absent when the pride ran from the Loliondo males; he was not around when Shiba needed to mate again. Lerai found her as she wandered alone and followed her patiently. She accepted him at last, weary of avoiding his advances. Some of the other Loliondo males tried to court the Sametu females but most of the females had cubs to protect and ran at the merest glimpse of the burly males.

The rains ceased. Slowly the plains dried up and the migratory prey drifted away westwards to the woodlands. As the plains emptied of prey, the predators of the plains began to re-sort themselves into different ranges. Hyenas moved back to the woodland fringes, cheetahs collected in areas where gazelles grazed, jackals and wild dogs found what space and prey they could and the lions reclaimed their dry season homes. The Boma pride left to return to its tributary of the Seronera river. The marsh and lake at Sametu were left open to the retreating herds and resident animals who welcomed the lion-free oasis.

The Sametu pride attempted to return to their valley, but found the Loliondo males still blocking them. The five males continued to range widely, patrolling the Sametu valley and adjoining ridges as well as retaining rights to the land further north and west. So the Sametu pride could not go home, nor could they go east to empty spaces where strange lions roamed over the sparse, dry grass. Nor could they go south where a resident pride and several tough adult males lived. In the chinks between these resident lions, the nomad trio ranged. The ridge of waterholes could not provide the large Sametu pride with sustenance to survive. When the water in the holes became a syrup of mossy mud, the lions began to shift northwards along the ridge towards the woodlands. Hunting was becoming very difficult and water almost nonexistent, and still they were chased by the Loliondo males. Sam got caught during one of the chases and received a deep bite in his thigh. He limped to keep up with the others and his sudden and belated appearance at mealtimes would gain him growls and snarls, for the females never knew any longer who would appear out of the darkness to try to join them.

Kesho stuck with the pride for a time but as conditions worsened he faded out of the scene: disappeared southwards to take up life as a nomad again. He could not compete with the wide ranging Loliondo males and gave up his throne and harem. He left the females and large cubs to defend themselves as best they could for he could offer no real help. So Kesho left his pride after almost three years – a long and productive time given the short life of a male lion. He left eleven young that were sired by either himself or his brother Kali. He also left several of the eight adults pregnant.

The pride was harried in its little corner of the plains and the lions grew thinner and more wary with each passing day. Moving northwards they eventually came to land that was totally unfamiliar to them. They hunted the few straggling gnu and zebra and tried unsuccessfully to catch stray eland, or even the fleet ostrich, for the area was poor in prey. The pride could not live there for long. They moved on towards the rim of trees on the horizon and found themselves in a different world. One morning they rested on a slope among whistling thorn trees. The wind had risen early and moaned across the openings in little black galls that crusted the stubby branches of the spindly acacias. A cub eyed the shrilling trees dubiously. Putting his nose too close he jumped away, sitting down on his brother behind him. It was not the thorns that had pricked him, but ants. A biting horde of small black ants had swarmed out of their gall homes at his touch, attacking his tender nose. The cub wiped his face and nose frantically then turned to his brother to rub his face along his

Whistling thorn

flank. At last the ants were wiped off and the cub moved away to rejoin the group.

Resting uncomfortably among scrubby grass, the lions looked down the slope to the thickening band of trees at the bottom. They had come to the end of Waterhole Ridge where it descended to a rocky river, hidden from them by a screen of contorted trees. To the east Zebra Valley met the river and to the west, the Sametu Valley wandered in. The thirsty lions lay panting through the drying day, smelling the water that was hidden from them by the trees. A bright moon kept the memory of the sun alive as they moved off in the evening, heading down the slope through the clumps of tufted grass towards the river. They went quietly across the moonswept slope and entered a world of strange shapes, sounds and smells. Cautiously ducking under the gnarled branches of the commiphora trees, they emerged through the screen of stumps. Before them was a stretch of grass. Pools glinted in the river-bed as it snaked through the valley, heading to a meeting with the Seronera river many turns away. The lions stood silently.

A scatter of black mounds became buffaloes in a stand of thick grass on the far side of the open stretch. Across the river a group of elephants moved silently into palm thickets. A herd of impalas watched the lions as they came through the trees to sit staring at the silent shapes. With graceful leaps and bounds, the impalas left the riverbank to disappear among the bushes and trees. Under tall acacias stood several giraffes, also spying the lions across the river. Silently the giraffes drifted away.

Overleaf: They entered a world of strange shapes, sounds and smells

Two dikdiks watched the lions from close by, hiding beneath thick bushes, their noses quivering. The pair held motionless as the lions came to the river's edge to drink. They lined the bank, a tawny ruffle on the edge of a silver ribbon. Drinking long and deeply the lions lapped the murky water. It tasted of soda but

Dikdik

was refreshing. The cubs finished and began to explore. They discovered aloes with bayonet leaves and tall flowers blooming for the sleeping sunbirds, picked up rocks and branches, carrying them about like meaty bones. A genet cat, surprised while crossing an open space, fled into a tall, leaning tree that hung over a bend in the river. A cub went up the tree trunk, discovering the wondrous fun of tree climbing. He was soon joined by the others; even limping Sam and lumbering Su tried to climb. While the cubs learned about falling out of trees, the adult females explored the river bank. There were vague scents of lion everywhere.

The females rested uneasily beside the water as their moon shadows grew shorter and the night longer. Unseen eyes watched them warily from high and low and sensitive noses caught their drifting scent. The pride went wonderingly along the river, exploring, sniffing, testing, discovering. Even the adults climbed or stood on limbs and rocks, surveying this moonlit world. The pride wandered westwards, following the river as it wended its way among rocks, trees and long grassy stretches laced by boggy patches.

Without warning the roars of several lions resounded along the river. With no hesitation the Sametu lions ran. They fled across the grass and through the rocks and aloes as a second fusillade of roars bombarded them. Back in the open grassland, they paused to catch their breaths and scan the moon-washed, open

164

Genet

sweep of plain as another onslaught of roars was hurled at them from behind the curtain of trees. The Sametu pride listened carefully, then moved rapidly on, back up the gentle slope and out on to the wide rolling plains. They stopped again to listen to roars coming from the south, far away in the direction of Sametu. The quiet, worried females stood undecided, then continued through the patches of prickly scrub, clumps of herbs and across the withering grass.

The Sametu lions had visited all corners of their range, tried to extend their roaming into new areas, and had been blocked on all sides. They had no choice but to continue living on the plains they knew, wandering like nomads, fighting to retain some rights to their disputed territory. They would need all their wits and luck to survive the dangers that dwelt there; dangers they would meet often during the menacing dry season.

CHAPTER TEN

Dispersed and Dispossessed

(Dry season, 4th Year)

Sonara accidentally discovered the ostrich just before dawn. She never would have caught it if the night hadn't been so dark and the ostrich fast asleep, sitting like a pile of stones, with its long neck laid out on the ground. The unnoticeable grey heap erupted into a flurry of feathers as Sonara passed close, startling it. With sure instinct Sonara caught the thin neck in her jaws. She hung on while the big bird thrashed about, trying to get its sharp toes into the lion's belly. After the ostrich became inert, Sonara let its neck drop gently to the ground and began to sniff it carefully. She had never caught an ostrich before and did not know quite what to do next. But she was very hungry and began to nibble at a thigh, trying to avoid the dusty, tickling feathers which clung to the carcass.

The stringy meat and fragile bones of the ostrich made a good meal; it was well after sunrise before Sonara had eaten her fill and abandoned the wafting pile of grey-brown feathers. A large white-headed vulture and a tawny eagle landed nearby and waited until Sonara moved off. Before the avian scavengers could investigate the remains, the two silver-backed jackals, who had been waiting since dawn, rushed in to snap up the remaining bones. The big vulture and small eagle looked disdainfully at the array of feathers but began to rummage through them anyway, looking for scraps.

Sonara walked away without so much as a quick glance behind her. Ahead, she had seen a vague shape that walked like a female lion through the long grass. Sonara moved quickly towards it, calling softly, eager to determine whether the shape was one of her beloved kin. These days she was often alone, and always eager for company. Last night she had been with Swala when they had been chased yet again by some Loliondo males. Running from the fourth

hill across the valley the two females had become separated, then Sonara had found the ostrich.

The shape on the slope was not Swala, nor even a lion. Sonara stopped as the warthog spied her and fled. Ambling along with only just its back showing, it had looked rather like a lion moving through grass, but when it ran, the bobbing rump and raised tail immediately marked the prey from the predator. Sonara sat down. She was tired, hungry, thirsty and lonely. Back in a brushy hollow below the fourth hill she had a litter of cubs she should return to and feed. She and Swala had had to bear their litters in these unsatisfactory dens since there was nowhere else to go. The last of Kesho's offspring, these cubs were doomed to be devoured by hyenas, killed by the Loliondo males, or abandoned by mothers who could no longer find time or energy to return and feed them. Unable to predict the future, Sonara and Swala cared for their litters as best they could, trekking back and forth to the helpless cubs despite the dangers. Sonara panted in the growing heat, trying to make up her mind what

White-headed vulture, silver-backed jackal, tawny eagle

to do next; whether to go all the way to Sametu for a much needed drink, to continue in hopes of finding friends or food, or to return to her cubs in the hot hollow.

The problem was decided for her by a loud bellowing from behind. Sonara glanced back to see two male lions lumbering up the slope from where she had left the feathers of her meal. She ran. Leo and Lengai began to roar at Sonara, trotting along shoulder to shoulder. They had been on top of the fourth hill when they had seen the vulture drop into the valley below. Running to see if the prize was worth taking they had found no scraps, only feathers and the scent of Sonara. Hardly pausing, Leo and Lengai had seen Sonara on the eastward slope and continued, following her. Well fed by scavenging carcasses from fleeing females, the healthy old males chased after thin, lactating Sonara.

Sonara headed south, towards the drier open plains where she knew the Loliondo males did not care to go. She ran fast, but not at full gallop, for she had learned she must harbour her strength and hoped the heavy males would tire in the heat. They soon did, letting Sonara go, stopping to roar, then walking up the slope where termite mounds adorned a barren spot, the den and home of many hyenas when the whole clan was there during the long rains. A few hardy dry-season hyenas slunk away as the big male lions took their termite mound thrones, scanning the undulating plains.

Sonara slowed her pace to a walk. She was very thirsty, yet the distance to Sametu and water was great. Also, the lake, wells and marsh were no haven, for the three nomads had claimed it. After the Boma pride had left and finally the Loliondo males as well, the nomad trio had moved in. For the three drifters, the place was a superb oasis, something worth fighting for. Nafasi and Nasibu readily defended their new domain from Sametu females and even Loliondo males. Bony Nani reaped some benefit from having a home base and secure hunting range and actually looked healthier than she had in years.

Sonara paused, panting hard, her tiny black shadow hiding underneath her for the blazing sun was now high. Lying down wearily she gazed back across the valley, then up the rise to the long ridge dividing Sametu from the Seronera watersheds. Where to go? She stood again and plodded down the slope, keeping

well away from the rise where Leo and Lengai rested. A loud humph from behind made her whirl, ready to flee or fight. Shiba stood still for a moment of recognition, then ran to greet Sonara. The two females rubbed against one another, then collapsed into the grass. They slept through the hot afternoon and awoke to a dusky, glowing sky.

The glows didn't come from the west but from the north-east, an odd place for a bright rosy hue. Rising to their feet, the lions could smell the reason for the glow more easily than see it. A large grass fire had started in the woodlands and was burning merrily. In the calmness of the evening, when the undecided wind eddied and flowed this way and that over the plains, the fire backtracked to devour its missed spots, burned its way across open stretches of grass and chewed into bushes and trees. Ashes and smoke spread out in the air far beyond the insistent flames.

The fire was a long way from Sonara and Shiba who were soon out of sight of its glow as they headed down the long rise towards the last of the four hills. They stalked gazelles, then kongoni, both to no avail. Going on again the two females came across the scent trail of other Sametu members. Crossing through a gulley and over another rise they found Sarabi and Swala sitting together in the starlight. After greeting, the foursome went south, hunting again the elusive prey. The glow from the north spread more widely as the great fire reached the edge of the woodlands and began to straggle out on to the plains, free to roam at will before the morning winds would wake to drive it westwards again.

Hare

Sarabi flushed a hare hidden in the grass. It ran, zigzagged and doubled back; Swala caught it, carrying it away to a grass clump to munch while the others sat on the exposed slope of a ridge leading nowhere, lost among so many other smooth and featureless ridges upon the wide plains. Sonara was restless, hungrier than the others and also eager to return to her cubs, two nights now without milk. She set out before Swala had fully eaten her small morsel, and Swala finished hurriedly, moving fast to catch up. The four headed towards the north-eastern rim of glowing gold, outlining the distant silhouette of Bustard Hill. Stalking and walking, they made their way along the flank of Ratel Ridge, as roars came from the trio of nomads at Sametu.

Roars also came from the Big Marsh where three Loliondo male voices rent the air. Taking a route between the two threats, the females found themselves at dawn on the slope above Warthog Valley. Dawn and the glow from the widespread fire blended to gild the coats of the weary lions. They were thin after months of running, hiding, hunting. Having to support cubs was an added burden to Swala and Sonara, who showed it by the roughness of their fur, clusters of ticks, protruding ribs and their ever-hungry look. Sonara sat on a termite mound to scan Warthog Valley. Across the valley on Bustard Hill an ostrich began to boom. Three measured throbs of sound drummed into the golden, hazy morning; "boom, boom, boooom". Had the booms gone on, the ostrich could easily have been mistaken for a lion roaring, so similar were its tones. Active ostriches were no lure for the lions; their sights were on a trio of warthogs that grazed openly near Warthog Rocks.

Drifts of ash and smoke from the spreading fire were collected by the morning breeze which swept the debris together and sent it in great plumes

westwards. In the clear air, the four lions began to stalk the warthogs. Down the slope they crept, moving quietly from clump to clump, holding still for long periods as the warthogs grazed into the increasingly strong wind. As the kneeling warthogs nibbled nearer to the rocks at the upper end of the valley, the lions carefully moved into position. Sonara was ahead of the others, keeping behind the last warthog. Swala and Shiba were in correct positions flanking the grazing group, while Sarabi moved around slowly to the side, ready to catch any hog that might run in her direction.

Sensing danger, the warthogs exploded into action, two sprinting away with Swala and Shiba in pursuit, while Sonara rushed the third. Two escaped, and Swala immediately turned to help Sonara who was still racing after her selected prey. The desperate warthog swerved, and Swala intercepted it,

grabbing its rump with her claws. They disappeared in a cloud of dust, just as Shiba caught up. The warthog screamed with its final breath before Shiba clamped her jaws on its throat. Sarabi and Sonara joined, panting, and soon all four were rapidly and quietly devouring the first real meal they had had in many days.

The strong wind and sun swept across the plains, as the females finished eating, then panted their way over to Warthog Rocks. Sonara carried the skull and hid in the lee of the rocks, while she gnawed the last scraps off it. Swala rested awhile, then reluctantly left to go back to her cubs. Though she waited and called, no one followed her, so she finally moved off on the long midday trek alone, Sonara was exhausted and slept soundly in the shadow of the biggest of Warthog Rocks. Over her watched a roller bird, sitting on a little twig, the only roosting spot in the valley. The beautiful turquoise bird flew down from time to time to catch insects deftly, returning to sit on his branch. In the afternoon he left to fly north, where the fire was burning more brightly now, beginning to drive out the insects living among the grasses on the plains.

The haze was thick and ashes fell in a dark gauze curtain over the grasslands when the wind dropped in the evening, letting the fire roam free again. A great crescent of flame flickered to life as the sky darkened and the fire crawled through the unburnt grass north of Bustard Hill. Sonara woke, nuzzled Shiba and Sarabi, then turned to leave, going up the slope from the rocks towards the rise that led back to the four hills and the distant den where her cubs slept. She dreaded leaving her companions to trek back alone but soon came across the scent of Swala who had gone the same way earlier, and felt vaguely reassured. As she walked in the darkness she passed some other lions, but she did not see them, nor they her. Sonara went on to her cubs in the gulley while Sam, Su and the four large cubs, watched three other lions intently. The six hungry youngsters were watching Sisi, Siku and Sega hunting a lone topi.

No topi was caught and the cubs rushed to greet the unsuccessful hunters as they came back across the slope. Turning away from the scene of yet another defeat, the hunters led the dependent flock up the rise. Roars came from the direction of the fourth hill and the entire group fled at once, going east, crossing Sonara's and Swala's trails unnoticed. More roars sounded from the valley bottom behind them and the cubs kept on, though the older lions slowed. Lerai roared again from the top of the fourth hill and was answered by Leo and Lengai by the Big Marsh. The wary members of the Sametu pride slowed but kept going south east, away from their ever present enemies.

The three sub-adults and four large cubs were all that now remained of the many cubs Kesho and Kali had sired in their tenure with the Sametu pride. It was a fair number of survivors, considering the problems of living on the plains. The young lions were lucky to have two adults with them, for they had already learned how hard it was to catch food on their own. Continually chased by the Loliondo males, the Sametu lions were often separated from one another. The cubs still tended to band together and tried to stick with the sub-adults who followed the fully adult and highly capable hunters. But it wasn't easy and many times the lions were scattered like dust motes over the plains.

Recently reunited, the lions kept close together, as they moved on steadily. Sisi and Sega caught a young topi calf that was hidden in the grass. Soon the small prey had been eaten and Siku and Su hung on to the remainder of the carcass. Su dragged Siku around as she pulled, trying to get the rest for herself. Su had lost much of the fat she had put on so easily during the good months of wet season. Her tail had slimmed to normal dimensions and her body was shaped less like a hippopotamus and more like a normal lion. Even so, she was still huge, larger than Siku or any of the other adult females. Sam was her only equal in size and he seldom contested her right to food, for Su was very possessive. The carcass parted and both Siku and Su got a portion. The others lay about waiting to go on, for neither their hunger nor their fear had been allayed. More roars came from the direction of the fourth hill, echoed by the booms of an ostrich to the north where the sky was vividly aglow with reddish smoke. The cubs trotted off as soon as the voices of Lengai and Leo came drifting from the valley bottom.

Escape of the aardvark

In a starlit patch of termite mounds a cub spied a moving hump and left the group to investigate. The aardvark moved more quickly as he approached. Suddenly the cub staggered backwards as he was caught in a choking cloud of dust. Coughing and sneezing he retreated, backing away from the frightened aardvark which threw clawsfull of dirt behind itself, rapidly digging into a hole and safety. Sam and another cub sniffed at the hole into which the aardvark disappeared. Their interest in this unique phenomenon waned when the sound of growls and chuckles came to them.

Rushing en masse to the scene, the lions found a hyena just killing a gazelle. More hyenas were appearing out of the dark and some lions tried to chase them away. In the mêlée, Su managed to get a solid hold on the gazelle, dragging it away from the others. Sam and Si sat near Su, hoping for a scrap, but they finally abandoned her and went on with the rest of the group. Turning north the lions trooped over the long ridge that divided the Sametu and Seronera valleys. Much later that night they reached Warthog Valley, empty of any lions but full of the fire's glow reflected off the blanket of smoke overhead. Over Ratel Ridge came roars from the nomad trio; down the valley to the west came an answer from two Loliondo males. The Sametu females and cubs listened. They never roared any more, moving always in stealth and quiet, listening to their foes and trying to exist in the limits of their shrunken range.

The ostrich on the flank of Bustard Hill started to boom as the morning wind began to bluster across the plains. The wind beat the fire into shape again, and pushed it westwards in fast running sheets of flame, clattering through the dry grass. The ostrich and lions seemed unaware or uninterested in the approaching fire. The ostrich was pulsing his pink tube of a throat, trying to impress his rivals and consorts, while the lions were merely weary, resting after the long night's trek.

The ostrich crossed the valley. He strode past the lions and on over the rise, beautiful in his black and white costume. He saw his female and slowed, booming three times, stepping along with delicacy through the brown grass, wings shuffling over his back. Up the slope from the lions was the blowsy female ostrich with her dusty grey-coloured wings. Seeing the male approach, she held her wings out from her body and waved them slightly. The male seemed to put his head up even higher, extending his already stretched neck, up and up.

The female ostrich kept her head down in a coy curve, holding her fluffy wings draped from her sides as if to dry them. She moved slowly up the slope. As the strutting male neared his lady love, he crouched and began to court her, swishing his wings back and forth, swaying from side to side, his pink neck puffed up. The lions watched the courting pair with interest as they would an entertainment, not as potential prey. The male ostrich swung his wings in great sweeps as he neared the female. She stopped, facing away from him, lowered herself into the grass, wings trembling. Her wings continued to quiver rhythmically as the male approached, holding white plumes over his back like sails. Stooping over his mate, still fanning backwards and forwards, he mounted her and began to copulate. His body swayed from side to side and his

Dance of the ostriches

neck began to sway too, curving and twisting like some fantastic snake. The writhing neck and sweeping wings moved to an inaudible ostrich mating rhythm, compelling music for the most beautiful dancers on the plain. The female ostrich held still, neck bowed, wings held out as they copulated. Abruptly she stood and moved away, holding her wings out, dragging them through the grass. The male stood, deflated his neck, ruffled and fitted his wings over his back, then headed down the slope, parallel to his mate. She would soon be able to lay more eggs in the nest she shared with his other mate. The bigamous male spotted his second spouse and turned to go up the rise after her going out of sight of the lions, into the growing red haze behind Bustard Hill.

Vultures dropping into Warthog Valley caught the lions' attention. They ran down the slope, hoping to find something to eat. At their approach, two Egyptian vultures flew away, followed more slowly by a tawny eagle and other vultures. Two silver-backed jackals turned to bark at the lions as they moved off, one carrying a white flake in its mouth. The lions found only the few bits of shell that were left by the squad of scavengers. The Egyptian vultures had found one lonely ostrich egg laid far from a suitable nest. They had broken the egg

Egyptian vultures

with a stone but had only got a beakful of the delicious liquid before the jackals had come, followed by the other vultures and the eagle. Ostrich eggs were a wonderful source of rich food that animals could only enjoy if they could figure out how to break into the big, smooth object, or had stolen it from someone who had already done so.

The lions sniffed the moist earth and a cub licked up some traces of egg, then they left the unrewarding spot, heading towards Warthog Rocks. There they found the welcome scent of recent occupancy by the other members of the Sametu pride. They also found the warthog skull that Sonara had left and two cubs chewed on it. Above them on the little bush, birds began to land, despite the presence of the lions. They were resting from their acrobatics in front of the advancing fire. The superheated wind swept the plains, drying grass and lions alike and driving the fire forward. Flames were wriggling along Bustard Hill and finally began to trickle down into Warthog Valley. Hordes of insects were flying and leaping and small animals ran from smoke-choked burrows. The kori bustard was at work, stepping among the darting flames picking up grasshoppers. A pair of secretary birds stalked along, catching lizards, insects and snakes. The tawny eagle found the hunting rich and took a mouse to eat on a termite mound. Jackals found a singed newborn gazelle and the roller bird flew in and out of the curtain of smoke catching the winged creatures trying to escape the flames. A marabou stork joined the throng, stomping along, clacking its big bill as it tossed roasted locusts down its scrawny throat.

The unfortunate lions could not share in the bounty gleaned from the onslaught of heat, smoke and fire. They reaped only thirst, as the thick ashes began to swirl over them in the dry, dry wind. Leaving the rocks they went up the slope, reaching the saddle on Ratel Ridge at dusk. They sat on the crest restlessly, looking down through the gloom into Sametu. The sounds from the fire reached them as it crackled, licking at the long, dry grass. Fiery streams flowed into Warthog Valley and dense smoke swirled above bushes or clumps that were wrapped in flame. The wind died, letting the fire lap its way across the grasslands, rivulets and cascades burning their way back over the ridge into Sametu.

Cautiously the band set off down the slope into the valley and could soon smell the water from the much shrunken lake. Everything seemed quiet at the marsh and kopjes as the group came near. They were almost at the shore when low roars came from Nafasi and Nasibu. Swallowing their thirst, the Sametu lions ran. Back along the slope they fled, Nafasi and Nasibu right behind. The two well-fed vigorous males loped along after the fleeing females and cubs, driving them away up Ratel Ridge.

From over the ridge to the west came roars from the Loliondo males. Hardly slackening their pace, the harried members of the Sametu pride curved around, running north into the glowing darkness. Necklaces of fire stretched across the plains around them; the blackened face of Bustard Hill stood stark and still ahead. The Sametu lions ran on through the flames and traceries of grey ash that adorned the smouldering slope; bare paws on hot black cinders and sharp burnt stubble.

Marabou stork (top), secretary bird and lilac-breasted rollers, hunting along firefront

CHAPTER ELEVEN

Reclaiming Sametu
(Wet season, 5th Year)

The fire roamed over the plains, grazing the old, tough grasses, but growing thin and straggling as the fodder became shorter in the south. By day the wind shepherded its flocks of flames with diligence, driving them towards the woodlands in the west where the fallen trees, many bushes and long grasses provided better pasture. At night the fiery herds strayed, flickering half-way into Sametu and crawling as far as the third hill. Then the droves were collected and pushed west again, crescents of fire creeping and consuming, leaving swathes of blackened earth behind.

Black dust devils danced over the barren landscape, some growing huge as they fed on the loose ash, funnelling high into the air and spiralling after the flames entering the western woodlands. The ash and smoke in the air stimulated the growth of storms, and rain came to Serengeti. Where red and yellow flames had flickered, bright green shoots of grass began to pierce the sooty stubble. Imprints of many hooves marked the fire and wind-swept slopes as animals trekked on to the plains. Food and water became plentiful for the scattered Sametu lions, but the problem of territory and males remained unsolved. Also, the young Sametu members had reached a turning point – their time as dependents was over, where were they to fit into the adult world?

Sisi sat alone among new grass and charred old clumps. Her gaze wandered over Warthog Valley, stopping at a group of topi, then passing on to a family of ostriches moving up the slope of Bustard Hill. The father ostrich looked more anaemic now, striding along with his sombre mate. She had laid final claim to his attentions, and together they had brooded a nest of over twenty eggs. So far, eight fluffy youngsters had survived the dangers that await eggs and chicks.

Their parents kept careful guard over the brood, leading them up the hill in a tight group. Sisi watched the ostriches go over the brow of the hill, then swung her gaze along the back of Ratel Ridge. The moon was just beginning to peek over the edge. From below came the voices of the nomads at Sametu. Sisi listened intently as Nafasi, Nasibu and Nani claimed the night. A counter-claim soon came from the west. Two Loliondo males roared from the Big Marsh and were answered by one of their brothers further south. Sisi absorbed this information while examining the hollows and curves of the slopes around her, looking for lion shapes.

Sisi caught sight of something in the sun's last wink, a subtle wrongness in the form of a termite mound. Actually, to Sisi it was a subtle rightness, for she recognised the distortion of tower and turrets to be due to a lion lying on top. She set out at once, stepping carefully over the prickly stubble that hurt her paws. Entering the tufted, unburnt grass, she moved more rapidly along the rise towards the moon, now sailing clear of Ratel Ridge.

Sisi had been following a scent trail so she knew that the lions silhouetted against the moon were friends. Nevertheless, she had learnt to move with caution, silently, alert to danger. Sonara saw her coming from her elevated position on the termite mound. She rose to her feet, eyeing the approaching Sisi carefully. Sisi slowed her pace, apprehensive about her welcome. The pride members, so often divided, were very wary; also, the adult females of the pride began to be more emphatic in their attempts to get their surviving offspring to

182

fend for themselves. Since all the cubs born during the past year had died, only the two sets of older offspring remained. The four large cubs were over eighteen months of age and had their adult teeth while Si, Sam and Su were two and a half years old, capable of surviving on their own.

The large cubs were now seldom accepted and Sam and Su had often been rejected as well; they were not very clever at keeping up with other pride members, so they were even less welcome when they did show up, usually at meals that they had not helped to secure. Sisi had not seen or smelled Sam or Su for many days, yet had crossed the scent trail of some of the large cubs. She missed her youthful companions but not as much as she missed the reassuring presence of her older kin with whom she could hunt and who were swift to discern danger.

Sonara met her with face rubs and a playful swat. Sisi slid against Sonara's smooth side, luxuriating in the contact with her friend and mother. Then she turned go greet Swala and Shiba, rubbing each furred face with care, reanointing herself with their scent. A cloud of feminine fragrance hovered over the little group as each of the three adult females rose and scraped the ground where she peed, ensuring that the scent covered feet, tails, soil and grass. They stretched, raking their claws along the ground, repeatedly tearing at the short clumps of grass. They seemed to mark even more deliberately than usual, not pausing as roars came from prides north and south. But they stood alert listening intently when roars came from the Loliondo males in the west followed by Nafasi and Nasibu at Sametu.

Sisi automatically turned south, assuming that the group would hunt in the space between the two dangers. But Swala had other ideas; she set off towards Sametu. Sonara and Shiba readily followed, Sisi coming more reluctantly. They

crossed the half-burnt saddle where the ratels rippled away, silver sheens over the dark, burnt earth. Entering the thick, unburnt grass again, the foursome came to a halt on the shoulder of Ratel Ridge. They lay among the clumps and looked into shining Sametu. The Sametu females had kept quiet ever since they had lost Kesho and had been on the run; roaring was a form of contact that was too revealing. But Swala voiced a low moan as she stood swishing her tail, setting off down the slope towards a spot moving to meet her, which proved to be Sega. After greetings, all drifted down, going more slowly. They stopped again to watch the three nomads descend the flat rock. Swala left her companions and went on alone.

Sonara took a few steps, then turned to look over her shoulder at Shiba, inviting her company before proceeding. Swala walked ahead, fearlessly approaching the three foes. The nomads watched her coming, staring at her in surprise as she stopped a short way from Nafasi. He understood her intent with amazing speed and stood to walk towards her. Swala twisted her head sideways, submissively, turned and walked away, flicking her tail repeatedly and pausing to scrape. Nafasi received these encouraging signs with deep inhalations, moving rapidly up the slope after the temptress.

Nasibu and Nani sat bemused, as if amazed that Nafasi could be won so easily by that brazen stranger. Then Nani noticed the other Sametu females on the slope, two of them coming down; she moved a little closer to Nasibu. Sonara and Shiba strode purposefully along, not hesitating until they came quite close to Nani. The bony female slunk sideways, trying to keep behind Nasibu. The stiff and stupid male finally rose to his feet, arching his neck, bewildered by the dance of defiance taking place around him. Nani slithered around to lie a short distance from her male companion. Sonara and Shiba followed her relentlessly, by-passing Nasibu and facing Nani squarely. Nani lowered her head, snarling at them.

Sonara and Shiba trotted along, forcing Nani further into nomad's land

Sonara and Shiba lay down opposite Nani, tense, heads lowered on to outstretched paws, ears erect, eyes narrowing, tail tassels vibrating to a rising rhythm of controlled fury. Nani broke first, uttering a deep-throated threat as she lunged at Sonara. Sonara met the attack with claws extended, teeth bared. They clashed, rolling over, grappling on the ground. Shiba stood ready to join the fray, leaning forward as if to spring, whiskers bristling. She leapt at Nani as Sonara rolled free, but Nani twisted aside, sprang to her feet and raced away. Sonara and Shiba chased her, driving the solitary female along the lake shore, through the thick grass around the sentinel acacia, and across the gradual incline to the south-east. Nani loped along, keeping just ahead of her pursuers. Neither she nor they had been hurt in the scuffle, but she knew their threats were serious.

Sonara and Shiba trotted along, forcing Nani further into nomads' land. Finally they slowed, walking along rubbing their faces together, nuzzling and reassuring each other. Confirming their unity, they stopped to roar, underlining with every bellow their right to be in Sametu. The forceful, triumphant duet rang out. Nani slowed, glanced behind, then continued into the lonely spaces. Sonara and Shiba turned, letting Nani go her way. Side by side they returned to Sametu, strode past Nasibu, but paused to scrape carefully beside the lake and marsh. Continuing between Jasmine and Fig Tree Kopjes they began to trot again, having heard the last bleat of a wildebeest. Sisi and Sega stood over the carcass of an old male gnu and shared the meat with their honourable defenders. Sarabi arrived out of the moonlit land to share too, wedging in with hardly a preliminary pause.

The meal was almost finished by the time Swala came, Nafasi trailing along as if on a leash. Sisi was alarmed at the approach of the strange male and moved away with a hiss and a moan. The others looked around them, seeking the

source of danger. They followed Sisi a little way but lay down again, cleaning their faces and paws while Sisi sat in distress. She tried repeatedly to get them to move further away but they lay here and there, in the moonlight, relaxed and unconcerned. Swala gleaned some meaty scraps from the bones while Nafasi sniffed all over the place, absorbed in the pool of scent. He inhaled deeply and often, and left his own mark repeatedly.

Sonara finished grooming herself and returned to the carcass, sliding under Nafasi's chin, leaning against his blond hairy chest, her tail curving and swishing boldly. Swala made no move to reclaim Nafasi's attention but kept on chewing. There was plenty of time for her to get what she wanted, and males were to be shared as much as food. Dawn came and slowly and singly, the older females left Sisi to walk back into the heart of Sametu. Sisi followed well behind, against her better judgement, very uncomfortable as waves of male scent wafted over her in the cool morning breeze. She stopped before reaching the kopjes and watched Sega settle herself near the lake. On a tier above reclined Shiba. Further up still were Sarabi and Nasibu. The playful, easy-going female had abducted the slow-witted male while her sisters slept. The others waited their turns. Nafasi could be heard growling lustily as he mated with Sonara.

Distressed and apprehensive, Sisi left the scene, heading alone up the slope of Ratel Ridge. Behind her the colours of dawn were reflected in the lake, and spread over the lions lying in ripples at the bottom of the slope. Sisi left the five voluptuous females and their two willing consorts, hoping to encounter any of the three remaining adults or even the large cubs who could still be scented in the area. A long pause to look around on Ratel Ridge did not reveal any lions so Sisi went on down into Warthog Valley, crossing no friendly scents en route.

She watched a distant line of wildebeest moving over the dark earth, flushed with new green grass. Storms had been increasingly frequent since the fire and many animals were returning to the plains. An experienced and adept hunter, Sisi could feed herself, but like all mature lions, she preferred to hunt with companions. That way she could eat more often as well as capture larger prey. Hunting with others was far more rewarding than scavenging food, as Sam and Susu so often did, and might still be doing, if they were still alive.

Wandering along she came upon the scent of Sukari and hurried to catch up. Sukari greeted her and the two sat together for a while as the sun rose. They watched the gnus trickling on to the plains and were startled by calls from behind them. Turning, they saw Siku trotting up and rose to welcome her. Behind Siku came an apprehensive large cub who did not dare to come close. He stood awkwardly, apparently afraid to ask for recognition. Sisi greeted him though, and he settled into the grass a good distance from the three older females.

Although it was broad daylight the females found the gnus irresistible as they straggled across the slopes, so intent on getting south to the short grass that they paid little heed to their surroundings. Moving downslope with the large cub trailing they found a better position among thick old grass that had escaped the fire, but the gnus they were watching fled as another lion approached them from the other side. Safi was walking along, heading steadily up the slope towards Ratel Ridge. Siku called and Safi stopped, glancing over to the immobile figures lying about in the grass. She immediately detoured to come to them, calling softly. As the females greeted one another, the male cub leapt to his feet and ran off to join a small shape trailing Safi at a discreet distance. It was his sister; the two cubs met with all the joy of long separated litter-mates, then lay down together at a distance from their elders. Sisi took care to greet Safi thoroughly, and also renewed her friendly gestures to Siku and Sukari. Then she greeted the two big cubs, letting them nuzzle and play briefly before returning to the adults.

The morning sun was now covered with damp cloudy blankets and the oppressive heat made the little remnant of the Sametu pride very sleepy. They stretched out among the grass clumps. Sisi remained awake, watching the splendid sight of more and more wildebeests galloping across the plains. Unknown to the little collection on the slope, some well-known but unloved individuals lay just a hyenas' lope away up the slope. Three Loliondo males were having their own reunion out of sight of the victims they had spent so many months chasing.

Leo and Lengai had come from the Big Marsh and had made a detour to meet Lerai. They met one another on the ridge crest where the local hyenas had their den. The black hyena cubs ducked into their burrows at the base of the termite mound as the lions passed by. The adult hyenas hunched off to watch from a careful distance. Leo and Lerai met with huge, shaggy head rubs. Lerai

then flopped down on Lengai who shoved his brother playfully. They circled around, leaning and sliding against each other. Lengai rolled in some zebra dung to enhance his smell, while Leo licked some droplets of water from Lerai's mane. It had begun to rain a little, just enough to make the grass sparkle and the air smell pungent and fresh. The old lions had well-worn faces full of character, they had lived long and nobly. Though they had two prides left, they had not relented in their ambition to take the Sametu females as well. The roars they heard from Sametu had warned them of Nafasi and Nasibu's presence as potential rivals, but that did not worry them at the moment; they stretched out in the light drizzle to sleep.

A larger storm was gathering by late afternoon, and the first fat drops woke lions here and there. Sisi and Sukari were already up and hunting before the deluge came. The Loliondo males woke fully when the rain poured down, darkening the horizons, laying their manes flat in dripping strands. They turned eastwards to listen to roars from Sametu. Flicking their tails over each other's backs, sliding and rubbing faces, they began to move in that direction, heads high. But a strangled cry followed by muted growls came from behind them through the falling rain. They turned back with another objective in mind, trotting swiftly past the hyena den.

Sisi and Sukari had caught an old gnu which had stumbled around in the rain and dark. The hungry group had just begun to eat, when the two young cubs bolted. Sisi, Sukari and Siku sat up, watched the retreating cubs, then spotted the cause. Three males were trotting down the slope towards them. The females ran too, separating, as loud roars blasted them. Safi stayed a moment longer, so very hungry, but plunged away just in time as Lerai pounded up. Lerai stood tall, watching her disappearing over the rise, while Leo and Lengai pranced up to the abandoned carcass and began to eat. Lerai took only a few bits then went on after the Sametu females, following Safi's tail, a relentless pursuit. For over a year the males had been after these females and they did not give up easily. Rain drizzled awhile on the matted manes of Leo and Lengai who were thoroughly muddy and gory by the time they had sated their appetite and departed in search of different but still desirable prey. Hyenas homed in on the remains, delighting in the free feast.

Sisi was far away by then, alone again. She wandered along the side of Warthog Valley, still hungry, for she had barely got a mouthful after releasing her stranglehold on their prey. No familiar scents wafted to her as she plopped along in the soft rain, her hide itching. She stopped to lick herself as the clouds pulled apart to let the moon shine down, making the grass sparkle. Sisi got a face full of drops as she nipped off some sweet green grass shoots. She wandered on in the moonlight, looking and sniffing but finding no one.

In the morning she was near Warthog Rocks and went to the tallest rock to lie in the drying breeze, scanning the valley, absently noting the ostrich family going along the flank of Bustard Hill, still intact, with all eight youngsters. Another family, of warthogs, grazed peacefully and several zebra families moved steadily over the rise crossing from the burnt area to the older but more abundant grasses. Topi and kongoni groups clumped here and there. All the world seemed to be in families, all except Sisi.

She slept the day away and woke early as rain pattered over her, sat up, stretched and departed. Walking across the valley and up the side of Bustard Hill, she saw neither friend nor foe. The rain ceased as she reached the rocky top, then walked slowly downhill, keeping to the ridgeback. She stopped to rest on the short grass where the kori bustard patrolled. A masked face popped out of a hole and peered at Sisi, then pulled back. When Sisi didn't move the little face came out again and after a long time a shy aardwolf emerged. It moved away stealthily, a shadowy, striped, thief-like animal. It looked more like a small hyena than a wolf, but unlike either it lived on termites. With sensitive

Aardwolf

ears it listened for the rustle and rattle of termites chewing up dead grass stems from inside their fragile earth tunnels on the soil surface. The aardwolf carefully kept clear of Sisi and moved away, nose low.

His passing seemed to wake the dreaming lion who rose and started downhill again, vaguely following the aardwolf. Reaching a large community of termite mounds she chose the biggest to perch on, trying to see and hear through the darkness. The aardwolf chose a mound too, but for a different purpose; he began to gobble up termites, catching them as they emerged from holes. Soon even Sisi became aware of the phenomenon that was so very

189

evident at the beginning of the rains: the exodus of winged termites from their colonies. The still, heavy air was filled with termites fluttering on papery wings all over the hill and plains. The aardwolf was enjoying a once-a-year banquet and smacked its lips as it crunched the tender morsels by the hundred.

Sisi was lying on an abandoned termite mound. Most of the structures were built over many years, then the builders would die out, leaving their monument standing. The deserted mounds with their deep air-vents made homes for many creatures: warthogs, aardvarks, pangolins, porcupines, mongooses, bat-eared foxes, jackals, hyenas, aardwolves, hedgehogs, lizards, and birds. Less than half of the termite mounds around Sisi were inhabited by their makers and these

were now sending forth colonizers in droves. A family of bat-eared foxes came running along the ridge, stopping to leap and snap at the flying insects. The adult foxes jumped and accurately caught their prey, their neat muzzles snapping each termite in mid-air. The four young foxes rushed about, playing, tumbling, licking up grounded termites. One leapt to catch an airborne insect and missed, falling awkwardly, collapsing on to its sibling and setting off another bout of play.

Bat-eared fox

Sisi felt increasingly lonely. She left the scene of food and frolic, heading south along the ridgecrest. The ratels moved across their saddle; they too were rushing to harvest the free food, abundant for so short a time. Two side-striped jackals passed on their way to join in the feasting. Sisi paused to listen to roars from just over the ridge. The voices of Siku, Sukari and Safi could be heard as they moved rapidly towards Sametu. Their call was answered by the rest of the Sametu females who still lounged around their males, enjoying the security of Sametu and the company of their lusty new companions. Nafasi and Nasibu roared their own welcome to the three females coming home.

Sisi hesitated, wanting to join them all, but too afraid. The sound of the males still made her intensely anxious, conflicting with her equally strong desire to join her family. She continued along the ridgeback, crossing the saddle and entering the long unburnt grass, wet and soggy from all the recent rain. The Loliondo males roared in the west and were answered by the full strength of the Sametu pride. Sisi longed to roar with her pride, hunt with them, lie among them during the long daylit hours, eat with their warm bodies pressed close to hers, lick their furry faces, just *be* with them, but the presence of Nafasi and Nasibu deterred her.

She came to another group of termite mounds and chose a tall and lonely

190

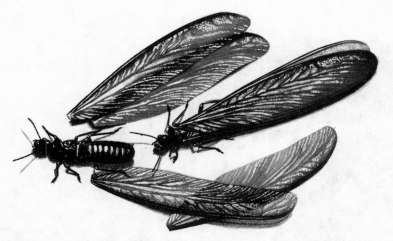

throne. Putting her head on her paws she listened to the silence that followed the roars. Just beyond her nose, a termite landed. The shiny brown female shrugged, and her four wings snapped neatly off at the base. Then she stood still, with her hind end raised, exuding scent. Soon a male landed and likewise shook off his wings. He began to court the female, following her nose-to-tail as she searched for a crevice. They would mate and stay together to found an empire underground, re-using the abandoned mound on which Sisi lay. Without noticing them, Sisi departed. She knew little enough about mating with *her* own kind. Her slender body was still growing but she was nearly adult. Deep within her, fluids and forces were gradually stirring to life. One day those frightening males would look different to her.

She walked further south along the ridge, leaving the widespread activity behind, going towards the heart of the plains where nomads ranged widely and resident prides were few and small. A hidden hare bounded away and Sisi deftly caught it, eating with relish for she was two days and nights already without a satisfying meal. The hare hardly filled her and she went on again but dawn soon came and hunting opportunities dissolved in the brightening sky. Sisi was by now at the last of the big patches of termite mounds before they petered out on the plains where there was too little grass, soil or moisture for the mound builders. She sat on a strangely shaped old mound, hollowed out by many different sets of claws, paws and hooves.

The termites had mostly landed at dawn and scurried wingless and vulnerable into any hole where they could escape the light. An early-rising group of banded mongooses was out, moving along like a single loose-jointed body, catching the stragglers, digging up termites half hidden, poking their noses under clumps of grass. The mongooses moved quickly from one spot to another, keeping in sight of each other, twittering constantly. Sisi sat up, half intending to chase them, thinking they might be edible, but the wary, alert animals chittered and ran further off, pausing to stand erect like little posts, then running until they could no longer see the lion. They piled up in a heap of grizzled fur and stripes, marking one another's backs, reassuring each other

Banded mongooses

just like lions and so many other social animals. Had Sisi caught a mongoose she would have left it uneaten for the musky, foul taste would have caused her to retch.

Sisi stayed on her mound for a while, then moved on again after sunrise, reaching an isolated waterhole. Hyenas reclined at the other end of the shallow pond, lying happily in the mud. They watched the slim female lion lap a little from the muddy pool, then went back to sleep, their deep dark eyes closing as Sisi wandered off. Some young hyenas played around the adults, romping and jumping on one another. Sisi left them behind and walked along a rise covered in short scrubby herbs. Where was she to go? She had come upon no friendly scents, only vaguely threatening smells of strangers, all male.

It grew hotter as the sun slashed its way through clouds. Sisi tried to rest on an open rise, all familiar landmarks hazed with distance. She was lonely, hungry, frustrated. She had a choice to make. Should she continue roaming alone with the hope of eventually finding some of the other youngsters, or should she try to rejoin the pride with its disturbing new males? She slept restlessly all day.

A sopping evening wept over Sisi and the Serengeti. She sat in the rain in the middle of nowhere, not knowing where to go. Moaning softly to herself she set out in no particular direction. She trudged through a valley bottom and up a slope where two jackals trailed her, yapping, spoiling any chances of a hunt. One more item of information was processed by Sisi's brain: how hard it was to hunt alone, how easily solitary hunts could be ruined. She hurried away from the jackals, reached the top of the slope and peered into the rain. Some hyenas were chasing a group of gazelles and brought down a young one. Sisi ran at the hyenas, expecting them to drop the prey, but they didn't. Instead, they turned to growl at her. She hesitated, growled back, and rushed in, too hungry for caution. The hyenas might have defended their kill but two male lions came running up, causing both hyenas and Sisi to flee. The two young nomad males got the gazelle and Sisi realised that scavenging would not be easy for a loner either.

Spotted hyena

She turned back to Sametu. The rain slowed and ahead in the gloom stood a female lion. Sisi stopped, watching the other female breathlessly, hoping it was one of her friends. She called low and went forward cautiously. The lanky female looked at Sisi and loped away. It was already looking better, with a full belly and smooth coat. But Nani moved away quickly, as furtive as ever. A single nomadic female was never relaxed, always having to be on the defensive against strangers, fending off other lions and scavengers from kills, hiding her helpless young, repeatedly losing them.

Sisi had no doubts now and walked steadily back to Sametu, reaching the far end of the valley before dawn. Sonara and Swala met her there, welcomed her briefly, then continued their hunt. They were stalking a lone zebra, an old male who no longer had any family. Single animals did not thrive in the social world of Serengeti. Sisi joined the hunt and was soon growling over her share of the carcass, shoulder to shoulder with her closest companions.

When the males came for a share, Sisi moved away, sitting at a distance awkwardly, like a chastised cub. She followed this pattern for several cycles of the moon, resolving the deeper dilemma slowly and with the help of time. Although she had definitely chosen the life of the pride against that of a lonely nomad, and the older females accepted her, she could not yet accept the males. Eventually she grew used to their presence; she had to, for they were fully incorporated into the pride by then. They had mated with every adult female except Sisi and had taken up the appropriate roles of defenders of the pride range, roaring, patrolling, marking and chasing away intruders.

By the time the bishop birds were nesting again and the steinboks were easily hidden by the long flowering grasses around the kopjes, Sisi could finally bear to sit near Nafasi and Nasibu. She accepted them with a three-year old's virginal shyness. Moving off one night from Owl Rocks, Sisi slid against Sonara,

193

then stood still as Nafasi approached to sniff her. She flicked her tail in his face and moved off with the other females. Bounding along, Sisi and Sonara chased one another over the long grass. Sonara crashed into Swala and Sisi leapt over Sarabi's back. They played like cubs, Nafasi and Nasibu stolidly bringing up the rear.

The pair of males were happy too, though they showed it by less flamboyant gestures. No longer nomads, they now had a place in and with a pride. Their big paws padded confidently over the fine soil. The nomadic life was hardly better for males than for females, a hard life of fright, fight and flight. Life with the pride was not only more secure, but more worthwhile. Their mating would produce results in a few months and then they would have real progeny to defend.

The females dug a warthog out of its burrow, a new skill they had added to their repertoire. After their meal they assembled on and around several termite mounds and began to roar. Nafasi and Nasibu led and each of the nine adult females added their distinctive voices. The great chorus rang out into the valley and spread over the ridges to east and west. Two Loliondo males answered to the north-west, still edging into the Sametu range, but the Boma females roared with the rest of the old males in their delimited range north of Bustard Hill. Then one familiar and two unfamiliar voices came from the south. Nani roared with two new companions, two young nomad males she had recently acquired. They would learn a lot from the ageing Nani, as well as sharing her food. Already getting lean again, Nani roared with enthusiasm, she didn't sound oppressed and was no longer lonely.

The females dug a warthog out of its burrow

CHAPTER TWELVE

Encore
(Onset of wet season, 5th Year)

The dry season was coming to an end. Some of the Sametu females gave birth to new litters several weeks before the onset of the rains. Thus the cubs sired by Nafasi and Nasibu were sheltered in safe dens around Sametu during the worst time of the year. Since their range was secure, the mothers could return whenever they needed to suckle the cubs. By the time the little ones had become restless and hard to control, needing supplementary food, the storms were lurking around the wooded fringes of Serengeti. Then the mothers began to lead their cubs out of their dens to join the pride. Soon there would be plenty of meat for the rapidly growing cubs; that is, if they survived to eat it. And the prospects this time looked much better than in the past.

Sonara and Shiba led their combined brood of five cubs from Sametu to Warthog Rocks. It was a long expedition but one filled with novelty and excitement for the cubs. Reaching the rocks in the middle of the night they stopped to watch a white-tailed mongoose saunter away, its tail fluffed out, glowing in the gloom. It made a detour to rummage through dung-pats in search of beetles but ran when a cub charged at it. The mothers let the cubs ramble around the rocks while they guarded them. With the rising sun, Sonara and Shiba sought higher perches, leaving the cubs in the centre of the rock cluster where they curled close in the grassy nest.

Above them, slouched over the round rocks, Sonara and Shiba dozed, looking indolent and relaxed. But they woke immediately when a lion figure appeared at the head of the valley. It was Sarabi. Seeing her sisters sprawled on top of the rocks, she came to greet them. Sarabi had been on her way to her own litter of cubs over the ridge, but lingered with her sisters while the warm sun

White-tailed mongoose

sapped resolve and energy. The afternoon breeze became more stimulating as it breathed over the lions the sweet earth-tang of a coming storm. Sarabi rose, called repeatedly to her sisters and got no lively response. So she left them, striding along the flank of Bustard Hill, going north towards a curtain of deep grey cloud.

Sonara woke in the evening to the sound of a zebra braying from the top of Bustard Hill, see-sawing his barks across the listless plain. The sound and sight of the striped herd on the hill top was most welcome. The stallion barked again, and Sonara rolled on her tummy to watch the zebras move along the crest of the hill while a whoop came from a lone hyena somewhere beyond. The spreading storm began to moan, low and deep, then the moan changed to a wail as the clouds darkened. For several days storms had been racing across the northern woodlands. Zebras and wildebeests and all the other migrants were massing there, building like a tidal wave, ready to flood out on to the plains as soon as the storms broke over the rich southern grasslands. Daily the dramatic storms, like this one, fought to claim the plains. The lions waited while the storm marched westwards.

At last it moved past and the din lessened. Swala licked the stray drops of rain from the cubs and then from Shiba's head. The voices of Nafasi and Nasibu could now be heard roaring from Ratel Ridge. Sonara and Shiba answered their males and simultaneously queried where their sisters might be. From beyond Bustard Hill came some answers. Far less welcome a sound was a murmur from the north-west. Echoing the grumbles from the departing storm, three Loliondo males roared together. Still lurking on the edges of the Sametu pride's range, they remained a constant threat.

Sonara licked a vulnerable cub affectionately. It squirmed and rolled away, wanting to play with its litter-mates. While the cubs tumbled and wrestled, Sonara and Shiba watched the distant zebras and listened to Nafasi and Nasibu

198

roar again. They slowly set off up the valley, taking their time, waiting frequently for the cubs to catch up. Near the top of Ratel Ridge the two males reclined majestically, watching the band of cubs bounding along between the two mothers. The troop gradually approached the fathers.

Nafasi and Nasibu received their offspring with gentle tolerance. Nasibu let a cub clamber over his big paw on his way to examine Nafasi, who sniffed the daring cub. It rolled over on to its back, exposing its white tummy, and Nafasi nuzzled the cub briefly. The cub got to its feet clumsily, then jumped at his father's woolly mane but slid off with a thump. Nafasi nuzzled it again. Leaping away, the cub rejoined its brothers and sisters, in a tumble of wrestling fur.

Sonara and Shiba led the toddlers away down the slope to Warthog Rocks for the sky was growing light in the east. One adventurous cub hissed in alarm when a black and silver shape came and went before his startled eyes. Sonara assessed the danger but it was only one of the ratels on its way to some burrow on the ridge top where it could spend the coming day in peace.

Halfway back to the rocks, the explorers stopped as Swala approached. She greeted Sonara and Shiba warmly, sniffed one or two of the cubs, then went on to join Nafasi, for she was ready to mate again. As the courting couple departed, Nasibu joined the others who had turned again in the direction of Warthog

Rocks, only to be met by Sarabi and Sukari who had been following Swala. They greeted each other and settled among the tawny grasses, Nasibu going off to mark a grass clump thoroughly before collapsing as if exhausted.

In the sultry storm-brewing morning, the cubs grew hot and crawled under the thick clumps of grass, but the adults could only stretch out flat. Sarabi became restless as usual and decided to return to her cubs in the early afternoon. Her leaving woke the cubs who began to pester their mothers. The group moved to the semi-shade and breeze provided by the nearby rocks where Sonara and Shiba could escape the irritable mass of cubs. While the lions slept, doughy masses of cloud filled the sky and they awoke in the evening to the smell of coming rain.

Swala and Shiba left the rocks to go hunting, but the cubs would not be left so easily and followed them. The mothers lay with the cubs for a time and listened to roars from Nafasi and Swala who were on top of Bustard Hill. Nasibu raised his head to listen too, stood, stretched and moved off slowly into a mist of golden rain that was beginning to fall. Sukari followed him a short way but turned to see if Sonara and Shiba would accompany her. The two mothers were having a difficult time getting the cubs to remain in the safety of Warthog Rocks, repeatedly going back, calling the cubs, then trying to leave quickly. Each time the cubs scrambled after them.

Shiba took it upon herself to stay with the reluctant cubs while Sonara set out at once to join Sukari. The two females moved rapidly up the flank of Bustard Hill into the singing rain. A chorus of frogs and toads clinked, croaked, droned, gurgled and rattled from their puddles all over the plain. It was another

Croaking toad

miracle of the Serengeti that the frogs could so suddenly appear at the onset of the rains. They, like the grasses, seemed to know just when to burst into new life. Dry season had definitely ended and the grass could almost be felt thrusting forth beneath their paws as Sonara and Sukari strode along.

Near the top of Bustard Hill they saw Nafasi and Swala sitting together. Sonara and Sukari went on, aware that some zebras were beyond the hill in a saddle between Bustard Hill and the next one north, a hill that was a vague boundary between the Sametu and Boma pride ranges. They stopped for awhile below the next hill, looking into the steady downpour, listening carefully. Nasibu appeared out of the gloom, returning from some private investigation. He sat by himself and bowed his soggy head, hunched against the rain. Suddenly he reared up when a voice humphed behind him. Shiba trotted up, slid against Nasibu's wet bib, flicked her tail in his face and ran to join Sonara and Sukari. Clever Shiba, helped by the rain, had persuaded the cubs to stay in a hiding-place among Warthog Rocks. As soon as they had settled down, she had crept from the rocks and hurried after her sisters.

The three females sat together, waiting for the rain to ease off, listening to the frogs, the distant whoops of hyenas, roars of lions, scuffles of hooves on wet earth – the symphony of Serengeti at night-time. They were setting out to hunt when a muffled bark came to them; they ran across the hillside to find Siku and Sega holding a dying zebra. The five females began to eat immediately while Nasibu sauntered up and sat at a distance, politely. In reality, he was waiting for reinforcements. A single male had to fight for his share of any carcass and five hungry females were worthy opponents. Nasibu hoped Nafasi would come along soon, for two males were given a wider berth and could claim an earlier (and therefore a bigger) share.

Nasibu stood up as more females approached. He was going to get even less than he had hoped. The three newcomers came forward at a dignified pace, leading a train of six little cubs, all wet but bouncy. Sisi, Safi and Sarabi brought their cubs to the edge of the ring around the carcass, then wedged themselves between the others, letting the cubs clamber in where they could. The cubs climbed between and over the steaming bodies, wallowing and falling among the heaving furred waves that were their mothers, aunts and cousins. Their teeth were too small to do much with the meat but they lapped at the blood and chewed on anything tasty. The females grumbled at the cubs just as they did at each other, low murmurs of warning, punctuated with snarls and slaps as one crowded against the other.

Swala arrived, stood watching the group, then joined in, trying to squeeze into the mass of eight large and six small bodies. Hardly had she secured a mouthful before Nafasi teamed up with Nasibu and the males stomped forth to get what they could of the remains. Females flew in all directions, removing any portable joints, leaving the carcass to the cubs and their fathers. Nafasi and Nasibu tried to ignore the cubs which climbed all over, slipping and sliding among the bones, getting thoroughly smeared with blood and mud. The cubs were an agreeable but persistent bother.

Soon there was virtually nothing left. Nasibu left to clean himself and explore the hilltop. Only Nafasi and one fat cub remained at the bones while all the other lions drifted away, lying about, grooming, licking and enjoying each other's company. The rain had stopped and an early moon was uncovered in the sky, shining on the clean, curved teeth Nafasi was using to glean every last morsel of meat. The tiny teeth of the cub beside him gleamed in a simple row, with a long way to grow before they matched Nafasi's lethal daggers.

Nafasi and the fat cub stopped and turned as a commotion arose behind them. The thin moonlight outlined a male strutting across the top of the hill. The stranger moved confidently, aggressively, head up, more like an intruder than one who was among family and friends. He stopped a little way from the group. The females slowly rose from their scattered grass clumps and assembled before their foe. They had recognised him. It was Leo, lifetime enemy, scourge of their youth, murderer of their cubs, stealer of their food, pretender to the throne of Sametu. Leo had heard the growls of feeding lions, but no roars, and had come to investigate; only now did he realise his mistake. They rushed at him. He backed away but he was knocked over by the onslaught of bodies. The defiant females would no longer run; they attacked with fury, growling, hissing and clawing.

Leo quickly regained his feet, turned and fled down the hill, back the way he had come. Nasibu ran past the females, following Leo, roaring. Nafasi was on his way to join Nasibu. They loped along behind the old male, prepared to chase him back into his own territory. Roars glissaded down the wet slope from the Sametu females, encouraging their defenders, threatening the intruder. Nafasi and Nasibu roared as they ran, getting closer to the flagging Leo. The old male reached the bottom of the hill, but instead of retreating further into his own territory, turned to face his youthful pursuers. Baring his broken and worn teeth he made a final stand.

Overleaf: Nasibu attacks Leo from the rear, while Leo lunges at Nafasi

Leo swung a massive paw at Nasibu who dashed aside, the blow glancing off his shoulder, the claws just grazing him. Nafasi aimed a bite at Leo's back, his clean canines digging into the worn hide. Leo howled and slashed Nafasi across the face. The old male sat down as Nasibu bit him on the rump. He tucked his vulnerable hind parts beneath him as best he could and snarled at his two opponents who stood opposite him.

Nafasi and Nasibu left Leo, giving him the chance to escape. They strutted away to roar, turning in all directions, broadcasting their presence to any and all in the neighbourhood. Then they returned to Leo, who had not taken the opportunity to leave gracefully. He crouched, pulling his bottom even further under him and swivelled to face them. Half rising, Leo lunged at Nafasi, grappling, sinking his teeth into Nafasi's neck but getting only a huge mouthful of hairy mane. Nasibu raked his claws down Leo's exposed back while Leo and Nafasi clawed at each other's faces and shoulders. Leo twisted as Nasibu bit him in the loin, releasing his hold on Nafasi's neck, tearing away a wad of Nafasi's silky mane. Whirling around, pivoting on his protected rump, Leo turned to do battle with Nasibu. Shreds of mane wafted in the pale moonlight as Nafasi flew at Leo, knocking the old warrior over. Leo sat up and slashed Nafasi again but the younger male leapt sideways, only losing a little of his hide to Leo's outstretched claws.

Nafasi and Nasibu circled Leo, then left him a second time. They moved off to roar again, defenders of their first real pride, their consorts and cubs. For them, everything was at stake, their entire future. Leo was still crouching where they had left him when they returned. He panted in great choking rasps. Too exhausted or stubborn to move, he twisted to face the third and last attack. When the two young males left again, Leo finally did get to his feet and staggered away painfully. Nafasi and Nasibu roared again and again letting the old male go. They had taught the intruder a long overdue lesson. Their warning was clear, written in blood all over Leo's scruffy hide: Get out, leave us alone, don't come back!

Nafasi and Nasibu went back up the hill, heads high, triumphant. Leo limped away and soon met Lerai who had come running to find out what was happening. He passed Leo who crumpled on to the ground. Lerai went to the spot where his brother had been attacked. He sniffed all around, then roared. From on top of the hill came the answers of the united Sametu pride. Lerai turned away from that newly enforced boundary and returned to sit near the badly wounded Leo, who lay still, his head bowed. After a long while, Leo lifted his shaggy head and began to lick his wounds. He was lucky to be alive. Before morning he could walk alongside Lerai as they went back into the centre of their range, meeting Lengai en route. The three walked slowly. Maybe they would have to keep to their smaller range now. Fights were harder and harder to survive and the Sametu males had proved they would and could defend their rights.

Lerai passed Leo, who crumpled on the ground

The Sametu pride also returned to the centre of their range. Nafasi, Nasibu and Swala remained on top of Bustard Hill, surveying the plains. Wildebeest, zebras, and gazelles were trickling back over the grassland, and the kori bustard boomed into the dawn. Sonara and Shiba went back to Warthog Rocks to fetch their cubs and bring them to Sametu to join the others.

Not far from the Sametu kopje, Sisi stopped at a rain pool. Leaning over, she broke her reflection with her pink tongue. She lapped delicately and was joined by Sarabi and Safi. Their six cubs tiptoed along the edge of the water, shaking their feet distastefully. One sniffed the water and got it up his nose. He sat on the sand and wiped his muzzle with a little paw. Siku and Sukari joined for a drink too and the five females caused ripples to spread, cross and break. A brave cub stood at the water's edge, fascinated by the tiny waves that slapped the shore. Sega nuzzled it as she also joined the drinkers along the sand bar.

The group lay on the sand in the early morning sun and greeted Sonara and Shiba and the rest of the young cubs. Then they all straggled back to Sametu, cubs bounding along. Crossing the sandy, bare spot, the females and cubs milled along the old bright beads and new shoots of green grass. They ascended the grey rock and were greeted by a flock of lovebirds which shrieked and chattered in the branches of the old fig tree. Flying away in a bright green cloud they went to Jasmine Kopje as the lions climbed on to the rocky platform. Sisi looked up at the few pairs of lovebirds that still clambered around in the tree, pulling off fruits. The fig tree was crusted with tasty pendants and its new leaves

207

had a coppery sheen. Masses of new leaves grew from the branches of the young fig trees, providing a shady hedge for the lions.

A cub fell into a crevice full of rainwater and shook off drops all over his companions. Sonara licked it dry while the adult females spread themselves over the rocks, cubs rushing from one to the other. Sisi lay under the small fig trees, a cub playing with her tail. She swatted it playfully with her tuft and teased her playmate by swishing her tail in loops, making it hard to catch. As the lions settled down, the lovebirds returned to the fig tree above them, stuffing their vivid orange faces with figs, trilling and chattering all the time.

Sonara looked out over the marsh and saw the serval stalking there. Along the lake-shore the blacksmith plovers were busy chasing another pair while a broad raft of ducks paddled past the avocets and stilts and the pair of Egyptian geese preened themselves. A capped wheatear sang from a rock behind the lions and was answered by another territory-holder closer to Owl Rocks. The steinboks were grazing peacefully nearby, keeping one eye on the serval and the other on the lions, but the owls had their big golden eyes tightly shut against the morning light, asleep in their leafy bush.

Sarabi was the last to join the others sprawled over the top of Fig Tree Kopje. She rubbed her face against Sonara's then did the same to Sisi before lying down half on top of her. Sisi grunted and pulled her foot from under her ever-playful companion. Some cubs came to partake of Sarabi's milk. At last Sarabi had a chance to rear cubs of her own. All the cubs had been born at a good time of the year, the range was secure and duties were shared between five mothers, all friendly relatives. Maybe this time, more cubs would survive. But Sarabi and the others did not worry about it. At seven and a half they still had quite a few cub-bearing years ahead. Each could produce up to six cubs in a litter at any time of year. The environment would determine whether the cubs lived or died, and meanwhile the mothers would do their best to rear them.

Sarabi nuzzled a cub and closed her eyes. Sisi laid her head on Sarabi's rump and also went to sleep, two cubs at her nipples. Sonara remained awake a little longer, gazing at gazelles flickering along the ridge and zebras moving up the valley. Further south the sun shimmered on distant southern kopjes. Sonara and her sisters had never been that far into nomads' land, but that was where Nafasi and Nasibu had roamed and probably Nani and Kesho were there now. Also, out on those free plains another nomad group had recently taken up the roaming way of life. Sam, Su and three of the now sub-adult cubs had banded together, serving their apprenticeship in survival.

It has been four years since the nine maiden lions had set out together, in danger from the Loliondo males, to leave their home pride forever. They had accepted two males, lost them and gained two more, consistently rejecting the tyranny of their five old enemies, the Loliondo males. They had lost many cubs but those that survived were robust; one was still with the pride, an adult with cubs of her own, the others insecure nomads but alive and well. Sonara looked at a lizard stalking a fly across the rock, with a cub stalking the lizard. Fly and lizard both escaped, but the cub was caught by another cub. They chased each other over the kopje. Sonara laid her chin on her paw. She was at peace, lying among a close circle of friends and relations, the abundance of the plains stretching far and wide. The Sametu pride had nine adult females, plus two prime adult males and eleven cubs with a bright future. Not only that, they had Sametu, a very precious piece of the Serengeti plains.

LION NAMES

This is a complete list of all the lions named in the story. In addition there are four other named groups: the Masai (*mah*-sye), Boma (*bo*-mah), Nyamara (nia-*mah*-ra) and Seronera (say-ron-*nay*-rah).

The names are in Swahili except for a few in the Masai language. The Masai people gave names to most of the places in the story. The most important are Sametu (sah-*may*-too), Loliondo (lol-lee-*on*-doh) and Serengeti (seh-ren-*geh*-tee), the last referring to the open plains.

NAME	PRONUNCIATION	MEANING
Sametu Sisters		
Sonara	so-*nah*-rah	goldsmith, jeweller
Sarabi	sah-*rah*-bee	mirage
Shiba	*she*-bah	full
Safi	*sah*-fee	clean, neat
Swala	*swah*-lah	gazelle
Sega	*say*-gah	honeycomb
Siku	*see*-koo	day
Sukari	soo-*kah*-ree	sugar
Salama	sah-*lah*-mah	peace
Sametu Cubs		
Sam	sam	(short for Sametu)
Susu	*soo*-soo	hammock
Sisi	*see*-see	us

First Set Sametu Males
| Kesho | *kay*-show | tomorrow |
| Kali | *kah*-lee | fierce |

Loliondo Brothers
Leo	*lay*-oh	today
Lerai	leh-*rye*	fever tree (Masai)
Lengai	len-*guy*	god (Masai)
Lemuta	lay-*moo*-tah	(Masai place name)
Laibon	lye-*bon*	medicine man (Masai)

Nomad Trio
Nafasi	nah-*fah*-see	opportunity
Nasibu	nah-*see*-boo	luck
Nani	*nah*-nee	who?

CHRONOLOGY

A chapter-by-chapter account of the major events of the story are linked under the four groups or under individuals.

Chapter	Year	Sametu Sisters	Loliondo Males	Kesho/Kali	Nafasi
CHAPTER 1 *Sisters*	1971	Among 24 cubs born in the Masai pride are 9 females destined to become the founders of the Sametu pride.	Six young males (4 yrs. old) have left their home pride and have moved into neighbouring Seronera pride to mate.	Two sub-adult (2 yrs. old) males are still with their home pride.	Nafasi is born.
and		Growing up, learning limits of Masai pride range.	One of the brothers disappears		
CHAPTER 2 *Brothers*	1972	Fathers leave pride, are replaced by another pair of males who are in turn ousted by the team of 5 Loliondo males.	Extend interests to Masai pride, mate with females and cause young pride members to become peripheral or leave.	Leave pride, become nomads.	Growing up.
	1973	Mothers mating with Loliondo males; young males leave, 9 sisters become peripheral to pride.	Mate with females in both Seronera and Masai prides.	Join Nyamara pride, mate, father cubs.	Sub-adult in pride.
action begins here	1974 (1st Year)	Sametu sisters leave Masai pride area, are now reaching age of mating.	Now at peak of prime (7 yrs. old) range includes part of Seronera pride and all of Masai and Boma prides.	With Nyamara pride.	Leaves home pride and area

Chapter	Year	Sametu Sisters	Loliondo Males	Kesho/Kali	Nafasi
CHAPTER 3 *Sametu*	1975 (2nd Year)	Establish themselves as a pride in Sametu area.	Extend interests to Nyamara pride, retain part of Seronera and all of Masai and Boma prides.	With Sametu pride, mate with all 9 females.	Resides on plains, has temporary relationships with other nomads.
CHAPTER 4 *Cubs*	1975 (2nd Year)	Eight of the nine females bear their first litter of cubs in middle of wet season. When Sonara, Swala, Safi, Siku, Shiba, Sukari, Sega, and Salama bring their litters together there are 18 cubs.	Gradually leave Seronera pride completely. Stay primarily with Masai, Boma and Nyamara prides.	With Sametu pride.	Nomadic on plains.
CHAPTER 5 *Dry Season*	1976 (3rd Year)	Pride shifts range into Seronera valley. Cubs exposed to many dangers; by end of dry season only 3 remain: Sisi (Sonera's daughter) Susu (Swala's daughter) and Sam (Safi's son).	Patrol huge territory, encounter members of Sametu pride, probably kill some cubs in outer part of Masai pride range.	With Sametu pride but reluctant to go with females and cubs near Lolindo males' territory.	Retreats to woodlands with another nomad for duration of dry season.
CHAPTER 6 *Wet season*	1976 (3rd Year)	Pride centres activities in Sametu area. Siku, Sukari and Salama have new litters totaling nine cubs.	In conflict with another group of males. Relinquish Masai pride. Encroach on northern Sametu area with Boma pride.	With Sametu pride.	Returns to plains alone, meets Nani (mates) and Nasibu. They become the "nomad trio"
CHAPTER 7 *Hard Times*	1976 (3rd Year)	Salama has disappeared, her cubs are adopted by Siku and Sukari. Pride shifts range into Seronera Valley. Sarabi has cubs alone and loses all of them. Kali vanishes; Kesho stays with pride. Sega, Swala, Safi bear cubs at end of dry season, 9 in all.	Associate with Boma and Nyamara prides.	With Sametu pride. Kali vanishes, Kesho continues to mate with Sametu females.	Nomad trio does not leave plains. The 3 lions establish themselves in an area immediately south-east of Sametu pride.

Chapter	Year	Sametu Sisters	Loliondo Males	Kesho/Kali	Nafasi
CHAPTER 8 *Waterhole Ridge*	1977 (4th Year)	Pride extends range eastwards as migratory prey returns. Shiba has cubs by herself at Sametu. Boma pride and Loliondo males intrude into Sametu area.	With Boma pride the males move into Sametu area, begin to chase Sametu pride members.	Kesho only adult male with Sametu pride, avoids Loliondo males.	With Nasibu and Nani as part of Nomad Trio, maintain range south-east of Sametu.
CHAPTER 9 *Neighbours and Trespassers*	1977 (4th Year)	Shiba's cubs disappear in Sametu den. Kesho abandons pride. All the youngest cubs are lost as Sametu sisters and older cubs run from intruders.	Remain in Sametu pride area, chase Kesho and others except females willing to mate.	Kesho is pursued by Loliondo males. Abandons Sametu pride, leaves area to become a nomad again.	Nomad Trio defends range from intruding Sametu pride.
CHAPTER 10 *Dispersed and Dispossessed*	1977 (4th Year)	Sonara and Swala have cubs in Seronera valley. Lose all cubs. Pride splits up, all members run from any strange lions. Heart of Sametu area occupied by Nomad Trio.	Continue to pursue Sametu pride, maintain contact with Boma and Nyamara prides.	Nomadic on plains.	Nomad Trio based at Sametu kopje and marsh.
CHAPTER 11 *Reclaiming Sametu*	1978 (4th and 5th Years)	Sonara and Shiba chase off Nani. Sametu sisters mate with Nafasi and Nasibu. Younger lions leave pride.	Continue pursuit of Sametu pride members, but now encounter resistance from nomad males.	Kesho nomadic.	Nani is chased off by Sametu sisters. Nafasi and Nasibu mate with eight adult females. They father cubs, become active defenders of pride range.
CHAPTER 12 *Encore*	1978 (5th Year)	Sonara, Shiba, Safi, Sarabi, Sisi have at least eleven cubs between them.	The five old males (11 yrs. old) abandon contest for Sametu pride; retreat to Boma and Nyamara prides.	Kesho's fate unknown.	Nafasi and Nasibu fully established as resident males with Sametu pride.

ACKNOWLEDGEMENTS

We are grateful to the Government of the United Republic of Tanzania for preserving Serengeti as a National Park, and for permitting us to work there, to Derek Bryceson (the late Director of Tanzania National Parks), David Babu (Chief Park Warden) and all his staff, and Tumaini Mcharo (former Director of the Serengeti Research Institute) for their cooperation.

Our scientific study of lions was funded by the Science Research Council (United Kingdom) and the New York Zoological Society, both of which supported us throughout. We are also very grateful to our mentor, Professor Robert Hinde, for his administrative and general support.

We worked under the auspices of the Serengeti Research Institute, which provided us with a house and other facilities. Our own work was just part of SRI's long-term study of the Serengeti and its lion population. The Sametu story was given historical perspective by the work of our predecessors, Dr George Schaller and Dr Brian Bertram, who left us much valuable data and ideas. We are grateful to them and to our successors, Dr Craig Packer and Dr Anne Pusey, who kept us informed of the progress of the Sametu pride after we finished our study.

Our colleagues at SRI helped us in many ways; we would particularly like to thank George and Lory Frame who contributed many detailed and accurate sightings of our lions; Richard Bell and Hendrik Hoeck who piloted our lion-finding flights, and Edward Sichalwe who developed all our identification photos.

Nature Expeditions International, by sending us to Tanzania as natural history tour-leaders, provided us with useful opportunities to revisit Sametu and keep track of the lions.

This book was written and the pictures drawn mainly in the homes of Elisabeth Bygott, David and Grizel Wilkins, Brian and Kate Bertram, and Warren Hanby. We are immensely grateful to them all for their hospitality and forbearance.

Last and most important, thanks to the lions of Serengeti in general and Sametu in particular, who tolerated us with great dignity. Without their cooperation we would never have been able to share this book with you.

217

INDEX

Page numbers in **bold type** refer to illustrations.
Subjects such as "adoption", "aggression" etc. refer to lions unless otherwise stated.

220